ENGINEERING MECHANICS

Dr. Abdul Basit Saim

Toronto Academic Press

Engineering Mechanics

Dr. Abdul Basit Saim

Toronto Academic Press

224 Shoreacres Road

Burlington, ON L7L 2H2

Canada

www.tap-books.com

Email: orders@arclereducation.com

ISBN: 978-1-77469-780-1

ABOUT THE AUTHOR

"**Dr. Muhammad Abdul Basit** who is currently serving as Associate Professor at primarily did his Bachelors in Metallurgy and Materials Engineering from UET Lahore, Pakistan in 2006. Owing to his R&D experience in automotive manufacturing experience and academic career at Bahauddin Zakariya University, Multan, he was awarded MS-leading PhD scholarship by HEC, Pakistan, so he joined Nano-device Engineering Laboratory at Hanyang University South Korea in 2011. He is an expert in the area of fabrication processes for photo-electrochemical solar cells, screen-printing process for development of sensitizable photoanodes using various semi-conduction materials, various wet-chemistry techniques for development of quantum-dots and counter electrodes for photo-electrochemical solar cells. His area of specialization is quantum-dot-sensitized solar cells. He more than 100 international publications in reputed journals including various prestigious conference proceedings and has delivered many an invited talks at local and international forums. In addition, he has published multiple book chapters and covers articles (including Chemistry of Materials), while he has also been selected for a prestigious fully funded Postdoc-NeT 2017 Researchers Networking program by DAAD (German Academic Exchange Service) regarding the perspectives in German materials science and nanotechnology.

Contents

List of Figures

List of Tables

List of Abbreviations

CGS	Centimeter Gramme Second
DOF	Degree of Freedom
ER	Electro-Rheological
FEM	Finite Element Method
FPS	Foot Pound Second
MKS	Meter Kilogram Second
MR	Magnetorheological
ODE	Ordinary Differential Equation
PDEs	Partial Differential Equations
PZT	Lead Zirconate Titanate
SDOF	Single Degree of Freedom Systems
SMA	Shape Memory Alloys
VLT	Very Large Telescope

PREFACE

Engineering mechanics is a critical area of study for any aspiring engineer. It is the foundation of many engineering disciplines, providing a deep understanding of the fundamental principles that govern the behavior of materials and structures. This book is a comprehensive guide to engineering mechanics, covering a range of topics from the basics of statics and dynamics to more advanced concepts such as mechanical vibrations and virtual work.

The first chapter provides an introduction to engineering mechanics, exploring its historical development and relevance to modern engineering. It also introduces the fundamental concepts of forces, moments, and equilibrium, which form the basis of much of the subsequent material in the book.

The second chapter delves into the dynamics of rigid bodies, examining the motion of objects under the influence of forces and moments. This chapter covers topics such as the conservation of momentum and energy and the motion of systems of particles.

The third chapter focuses on statics and equilibrium, covering the analysis of forces and moments in stationary structures. This chapter includes topics such as free-body diagrams, analysis of trusses and frames, and the calculation of reaction forces and moments.

The fourth chapter explores the mechanics of materials, which is concerned with the study of the behavior of materials under load. This chapter covers topics such as stress and strain, Hooke's law, axial loading, and torsion. The fifth chapter introduces virtual work and energy, which is a powerful technique for analyzing complex mechanical systems. This chapter covers topics such as work and energy, potential energy, and the principle of virtual work.

The sixth chapter focuses on mechanical vibrations, which are a common phenomenon in many engineering applications. This chapter covers topics such as natural frequencies, damping, resonance, and forced vibrations.

The seventh chapter introduces simple machines, which are devices that enable the transfer of mechanical energy from one form to another. This chapter covers topics such as levers, pulleys, and gears.

The final chapter explores the applications of engineering mechanics, providing examples of how these principles can be applied to real-world engineering problems. This chapter covers topics such as structural analysis, machine design, and aerospace engineering.

This book is intended for students, scholars, and professionals in the field of engineering. It provides a comprehensive understanding of engineering mechanics' critical concepts and practices using a formal and informative writing style. It is a valuable resource for anyone seeking to deepen their understanding of this important field, and by reading this book, you will gain the knowledge and skills required to analyze and solve complex engineering problems with confidence.

—Author

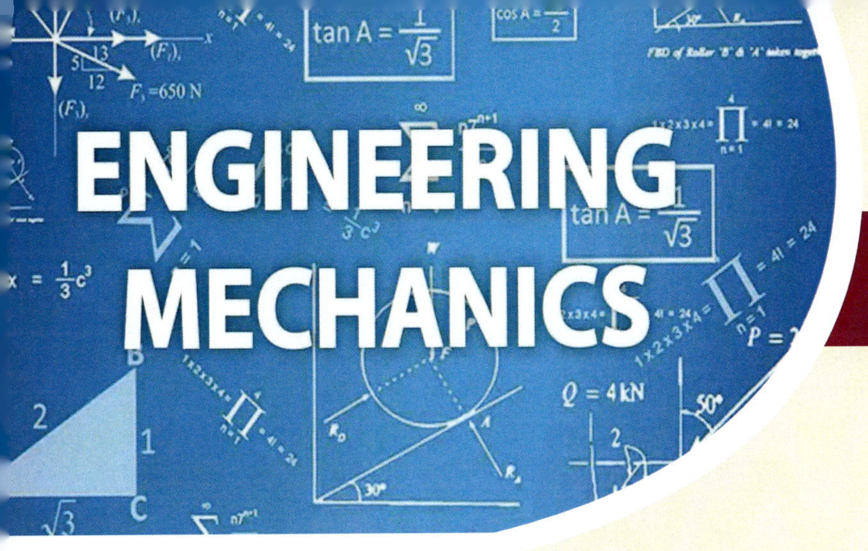

CHAPTER 1

INTRODUCTION TO ENGINEERING MECHANICS

UNIT INTRODUCTION

For many years, mathematicians, philosophers, and physicists have been interested in the states of rest and motion of objects under the influence of various forces. Mechanics is the branch of physical science concerned with motion or rest. The study of mechanics has evolved from analyzing solid bodies subjected to gravity forces and basic applied forces to studying robotics, airplanes, and spacecraft subject to atmospheric forces, temperature forces, dynamic forces, etc. The development of mechanics was greatly aided by Galileo (1564–1642), Sir Isaac Newton (1642–1727), Archimedes (287–212 BC), and Albert Einstein (1878–1955). Euler, Varignon, and Alembert all made significant contributions (Daniel et al., 2006).

These researchers' mechanics could be categorized into:

i. Newtonian mechanics/classical mechanics;
ii. Quantum mechanics/wave mechanics;
iii. Relativistic mechanics.

The father of mechanics, Sir Isaac Newton, combined the theories and experimental evidence gathered about the states of rest and motion of substances and presented them as three laws of motion and the law of gravitation. These laws form the foundation of classical mechanics, sometimes known as Newtonian mechanics. Albert Einstein demonstrated how Newtonian mechanics could not account for the behavior of fast (light-speed) moving objects. He proposed the relativistic mechanics' theory (McCormick, 2009).

Regarding atomic distances, Newtonian mechanics cannot demonstrate particle behavior. They presented the quantum mechanics theory. Engineers are eager to apply the principles of mechanics to real-world issues. Engineering mechanics is the branch of mechanics that applies laws of mechanics to real-world problems. Newtonian/Classical mechanics has withstood the test of time for all the issues ranging from atomic to high-speed distances. Hence engineers employ it. So, in this text, engineering problems are analyzed using classical mechanics (Muhanna et al., 2007).

Learning Objectives

At the end of this chapter, readers will be able to:

- Explain the importance of Engineering Mechanics in the study and application of mechanical engineering.
- Describe the historical development of Engineering Mechanics, including key contributors and milestones.
- Define and apply fundamental terms and concepts in mechanics, such as position, velocity, acceleration, and force.
- Discuss the Laws of Mechanics, including Newton's Laws, and their implications for mechanical systems.
- Identify and describe derived laws in mechanics, such as the Law of Conservation of Energy and the Law of Conservation of Momentum.
- Explain the role and importance of units in mechanics and convert between different unit systems.
- Identify and describe common features of force, such as magnitude, direction, and point of application.
- Analyze and calculate force systems using vector and scalar methods.
- Explain the importance of dimensional homogeneity in mechanics and use dimensional analysis to solve problems.
- Discuss the concept of idealization in mechanics and identify some common idealizations made in mechanical systems.
- Analyze the accuracy of calculations in mechanics and identify sources of error.
- Describe different approaches to solving problems in mechanics, including analytical, numerical, and experimental approaches.

Key Terms

- Acceleration
- Deformation
- Dynamics
- Elasticity
- Equilibrium
- Fluid mechanics
- Friction
- Kinematics
- Moment of inertia
- Statics
- Stress and strain
- Torque
- Vector analysis
- Vibration
- Work and energy

1.1. CATEGORIZATION OF ENGINEERING MECHANICS

Based upon the applied mechanics on the body, engineering mechanics is categorized as:

- Fluids mechanics;
- Solids mechanics.

The mechanics of deformable and rigid bodies are additional branches of solid mechanics. Rigid bodies are Newtonian ones that will not deform or in which deformation can be disregarded during analysis. Statics is the branch of rigid body mechanics that deals with bodies at rest, and dynamics is the branch that deals with bodies in motion. When dynamics deal with issues without mentioning the forces pushing a body's motion, it is referred to as kinematics; when it does so, it is referred to as kinetics (Bhatt & Rao, 2005).

The deformation of the body must be considered while analyzing the internal tensions generated in a body. Mechanics of Strength of Materials / Solid Mechanics/Deformable Bodies is the name of this branch of mechanics. The theory of Elasticity and the Theory of Plasticity are two subfields of this science (Mani et al., 2017) (Figure 1.1).

Figure 1.1. /llustration of strength of material.

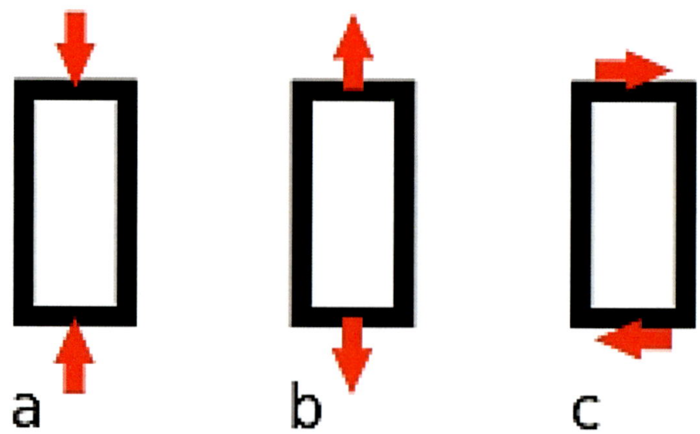

Source: Daniel M., Creative Commons License.

Continuous deformation of liquids and gasses is possible by applying very modest shear forces. These substances are known as fluids. Fluid mechanics is the branch of mechanics that deals with the behavior of these materials. Other classifications in this field include viscous fluid mechanics, ideal fluids mechanics, and incompressible fluids mechanics (Zhao et al., 2021).

1.2. IMPORTANCE OF ENGINEERING MECHANICS

Engineering mechanics is a primary branch of engineering that addresses how physical systems and objects behave when subjected to forces and motion. It offers a thorough grasp of mechanics' principles and their

use in mechanical, aerospace, civil, and biomedical engineering, among other engineering specialties. The following examples will help to illustrate the significance of engineering mechanics (Dubourg et al., 2013):

1.2.1. Design and Analysis of Structures

The theoretical underpinning for constructing and evaluating structures like bridges, buildings, and airplanes is provided by engineering mechanics. Structures' stability, strength, and load-bearing capability are assessed using mechanical principles. This information is crucial for ensuring that buildings are secure, resilient, and can survive harsh weather (Kirsch, 2003).

1.2.2. Optimization of Mechanical Systems

Improving mechanical systems, like motors, engines, and machines, requires a thorough understanding of mechanics. Engineers create and enhance the functionality of these systems to increase their effectiveness, dependability, and efficiency. They do this by using the principles of mechanics (Oakley et al., 1998).

1.2.3. Safety and Reliability

Engineering mechanics is essential to guarantee the security and dependability of engineering systems. The response of materials under diverse conditions is examined using mechanical concepts to assess their strength and foresee failure. Developing systems and structures that are secure, dependable, and capable of enduring harsh environments require using this expertise (Ching, 2009).

1.2.4. Innovation and Technological Advancements

The growth of innovations and breakthroughs has benefited greatly from studying mechanics. For instance, mechanical principles have enabled advancements in aerospace engineering, like constructing spacecraft and supersonic aircraft (Tatum, 1987).

1.2.5. Interdisciplinary Applications

Aerospace, mechanical, biomedical engineering, and Civil are just a few engineering specialties that can use mechanics concepts. Engineering mechanics is multidisciplinary, making it a useful tool for tackling crucial engineering issues and improving research and development across various disciplines (Stone et al., 2003).

To summarize, engineering mechanics is essential to many engineering disciplines because it provides a theoretical framework for building and assessing mechanical systems, improving them, maintaining their safety and dependability, stimulating innovation, and furthering multidisciplinary research. To build, grow, and innovate in their respective domains, engineers must have a solid grasp of the fundamentals of mechanics (Berthaume & Kramer, 2021).

Did you Know?

Archimedes, a Greek engineer in the 3rd century BC, developed the principle of buoyancy and invented the Archimedes screw, which are still used today in engineering.

1.3. HISTORICAL DEVELOPMENT OF ENGINEERING MECHANICS

A branch of engineering known as engineering mechanics examines how physical bodies and systems respond to forces and motion. Engineers, physicists, and mathematicians from diverse cultures and eras have contributed to developing and improving engineering mechanics. This article will look at the historical growth and development of engineering mechanics (Chen & Haik, 2002).

1.3.1. Ancient History

Engineering mechanics has its roots in prehistoric cultures, including the Egyptians, Greeks, and Chinese. Ancient mathematicians like Archimedes and Euclid created the statics and geometry principles as the basis for contemporary engineering mechanics. Archimedes is particularly renowned for his achievements in hydrostatics and the buoyancy concept (Tan & Tan, 2015).

The construction of enormous structures like the pyramids by the ancient Egyptians, on the contrary, required a sophisticated understanding of the response of materials under varied conditions. They did this by using mechanical principles. The Chinese developed the water clock and other mechanical innovations like the winnowing fan and wheelbarrow, who also made substantial contributions to the study of mechanics (Wang, 2005).

1.3.2. Renaissance and Enlightenment Periods

Engineering mechanics developed as a separate discipline of study throughout the Renaissance. Galileo Galilei developed the concept of inertia, and Leonardo da Vinci researched the mechanics of solids and fluids. Both scientists contributed significantly to the subject (Haas et al., 2007).

Engineering mechanics underwent significant advancements during the Enlightenment when mathematicians like Leonhard Euler and Isaac Newton created principles of calculus and dynamics that completely changed the industry. A comprehensive framework for analyzing the behavior of physical systems was supplied by Newton's law of universal gravitation and his three laws of motion; in contrast, Euler's equations of motion served as the cornerstone for contemporary fluid mechanics (Rose & Drake, 1971) (Figure 1.2).

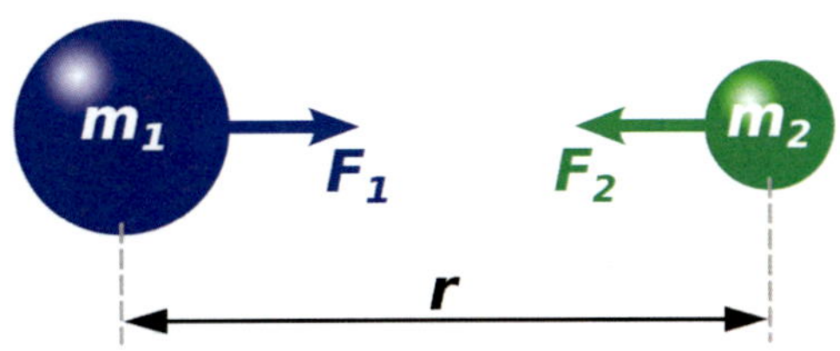

$$F_1 = F_2 = G\,\frac{m_1 \times m_2}{r^2}$$

Figure 1.2. Image of newton's law of universal gravitation.

Source: Dennis Nilsson, Creative Commons License.

1.3.3. Industrial Revolution

With the expansion of industries like transportation, construction, and manufacturing, engineers faced new difficulties during the Industrial Revolution, a crucial turning point in the growth of engineering mechanics. A detailed understanding of mechanics was necessary to create steam engines and other mechanical devices. In contrast, an extensive understanding of structural mechanics and materials science was necessary to construct large structures like tunnels and bridges (Vu & Le, 2019).

1.3.4. 20th Century

More developments in engineering mechanics were made during the 20th century, with new technology like cars, planes, and spaceships giving engineers new possibilities and constraints. Innovations like space shuttles and jet engines required detailed knowledge of thermodynamics, structural mechanics, and fluid mechanics. The mechanics' principles were subjected to these new technologies (Durst, 2000) (Figure 1.3).

Figure 1.3. *Top-down diagram of fluid mechanics.*

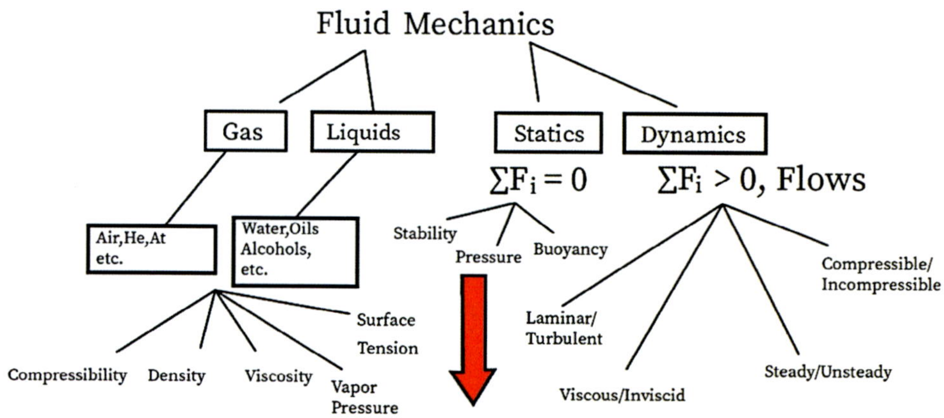

Source: Bartleby, Creative Commons License.

Engineers can now mimic the behavior of intricate systems and structures because of the advancement of computational techniques like finite element analysis, which was made possible by the advent of computer technology. Because of this, engineers can now build and improve systems and structures more effectively and precisely than ever (Barton, 1999).

Lastly, engineering mechanics has developed over the years owing to the contributions of engineers and scientists from diverse societies and eras. Engineers now face new difficulties and opportunities due to the growth of novel technologies and the expansion of industries, advancing the discipline. The cornerstone for resolving engineering problems continues to be the principles of engineering mechanics (Zingoni, 2001).

1.4. FUNDAMENTAL TERMS IN MECHANICS

The key terms for studying mechanics that you should understand completely are discussed in subsections.

1.4.1. Mass

The term "mass" refers to how much matter a body possesses. Unless a body is physically separated from itself or is harmed, its mass won't change. While a body's weight might change when removed from a spacecraft, its mass would not. This is because gravitational forces are subject to change. When gravitational pull disappears, even the body might be weightless while the mass remains constant (Lubarda & Hoger, 2002) (Figure 1.4).

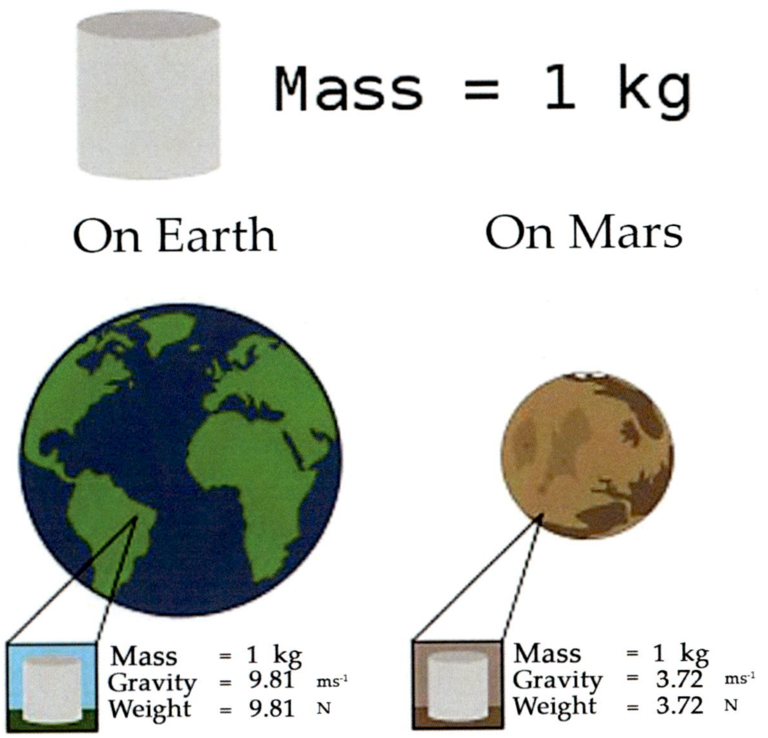

Figure 1.4. Illustration of mass.

Source: Master Uegly, Creative Commons License.

1.4.2. Time

The measurement of event recurrence is time. The next event chosen is the earth's rotation around its axis, known as a day. A day is divisible into 24 hours, an hour into 60 minutes, and a minute into 60 seconds to provide simple units for diverse activities. The tools created to measure time are clocks. The length of the cesium-133 atom's radiation over a period of 9,192,631,770 seconds is chosen as the time unit to get around problems caused by variations in the earth's rotation (Saunders, 1998).

1.4.3. Space

Space refers to the geometric area that includes body study. A series of angular and linear measurements can refer to a point in space to a present point. The origin and a set of measurements are called the reference point's "coordinates." Cartesian coordinates are those in which only mutually perpendicular directions are involved. The polar coordinate system is used when the coordinates include distances and angles (Feynman, 1948).

1.4.4. Length

It is a theory for calculating linear distances. A cylinder's diameter could be 300 mm, while a building's height could be 15 meters. The actual unit of length is the meter. Nevertheless, kilo, milli, or micrometer units are employed for measurement according to the concerned sizes. The International Bureau of Weights and Measures defines a meter as the platinum-iridium standard bar length. Today, a meter is defined as having a wavelength of 1690763.73 krypton-86 atoms to address availability and reproduction challenges (Takagi & Ichihara, 2004).

1.4.5. Displacement

The distance *x* that a body or particle moves in a particular direction is called displacement. On the x-y plane depicted in Figure 1.5, if a body travels from position A to B, its displacement in the x-direction will be AB′, and in the y-direction will be B′B (Burdekin & Stone, 1966).

Figure 1.5. Diagram of displacement.

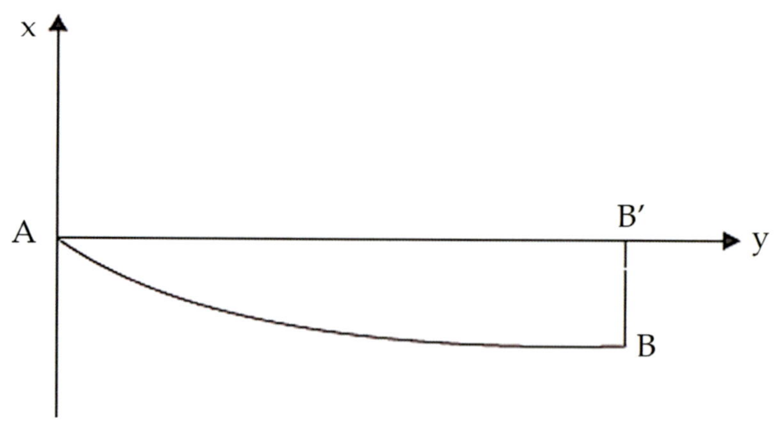

Source: Bhavikatti S., Creative Commons License.

1.4.6. Velocity

Velocity is the rate at which a given displacement changes over time (Young & Choice, 2007).

1.4.7. Acceleration

The rate at which velocity changes concerning time is known as acceleration (Morin et al., 2015). Thus:

$$a = \frac{dv}{dt}, \text{ where } v \text{ is velocity}$$

1.4.8. Momentum

Momentum is the product of velocity and mass (Edmonds, 1996). Thus:

Momentum = Mass × Velocity

1.4.9. Continuum

A body is made up of various substances. Each particle may further be broken down into molecules, atoms, and electrons, a well-known fact. Any engineering challenge cannot be solved by thinking of a body as a set of such distinct particles. The assumption is that matter is distributed continuously throughout the body. To put it another way, the body is considered a continuum (Kachanov, 1986).

1.4.10. Rigid Body

The body is considered rigid if the relative locations of any two particles within a body remain unchanged while forces are applied. Points A and B in Figure 1.6(a) represent a body's initial position.

When a series of forces F_1, F_2, and F_3 is applied, the body assumes the position seen in Figure 1.6(b). The new positions of A and B are A′ and B′. If the body is thought of as rigid, then A′B′ and AB are in the same relative position or A′B′ = AB.

Many engineering issues can be satisfactorily resolved by supposing that bodies are rigid (Peters et al., 1983).

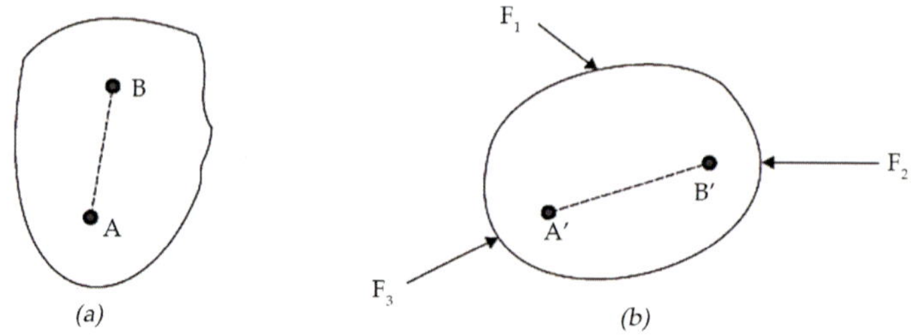

Figure 1.6. *Schematic of rigid body.*

Source: Bhavikatti S., Creative Commons License.

1.4.11. Particle

An object with mass and no size is referred to as a particle. Theoretically, one such body cannot exist. Even so, the body can be considered a particle for solving issues requiring distances significantly greater than the size of the body without compromising accuracy. Examples of these cases include: a bomber plane is an aircraft a gunner may operate from the ground.

A ship in the center of the ocean is a particle when examining the relative motion of a ship from a control tower. While examining the movement of the earth in the celestial sphere, the earth is seen as a particle (Nugent & Posch, 2000).

1.5. LAWS OF MECHANICS

The basic laws of mechanics are as follows:

- Newton's law of gravitation;
- Law of transmissibility of forces;
- Parallelogram law of forces;
- Newton's first law;
- Newton's second law;
- Newton's third law.

1.5.1. Newton's First Law

According to this statement, everybody remains in its position of rest or uniform motion in a straight line until forced to do otherwise by an outside force operating on them. As a result, force is defined as an external force that modifies or tends to modify the body's resting state or uniform linear motion (Galili & Tseitlin, 2003) (Figure 1.7).

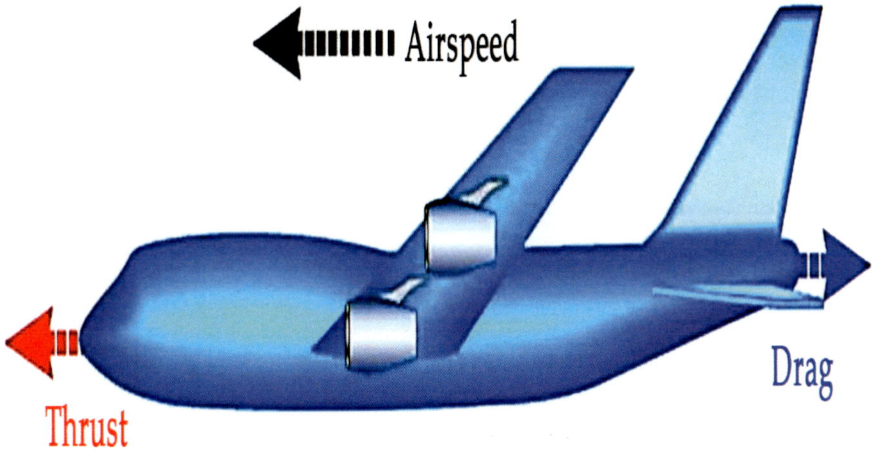

Figure 1.7. Example of Newton's first law of motion.

Source: Wondrium Daily, Creative Commons License.

1.5.2. Newton's Second Law

According to this, a body's rate of change in momentum is directly proportional to the impressed force and moves in the direction of that force. As a result, under this law:

Momentum = Mass × Velocity

Force ∝ rate of change of momentum

As mass does not change,

i.e., Force ∝ mass × acceleration

Force ∝ mass × rate of change of velocity

$F \propto m \times a$ (1.3)

Remember

First law challenges our intuition and is fundamental to understanding the behavior of objects in motion in physics and engineering.

1.5.3. Newton's Third Law

It asserts that there is an equal and opposite reaction to every action. Think about the two bodies close to one another. Let's say that one body exerts force F on another. This law states that the second body produces a reactive force, R, which is equal to force F in magnitude and exerts in the same direction as F, but in the opposite direction (Pourciau, 2006). Figure 1.8 depicts the movement of the ball and the ground's response. The motion of the ladder on the wall and the floor, as well as the responses from the wall and floor, are depicted in Figure 1.8 (Poncet & Bartolo, 2022).

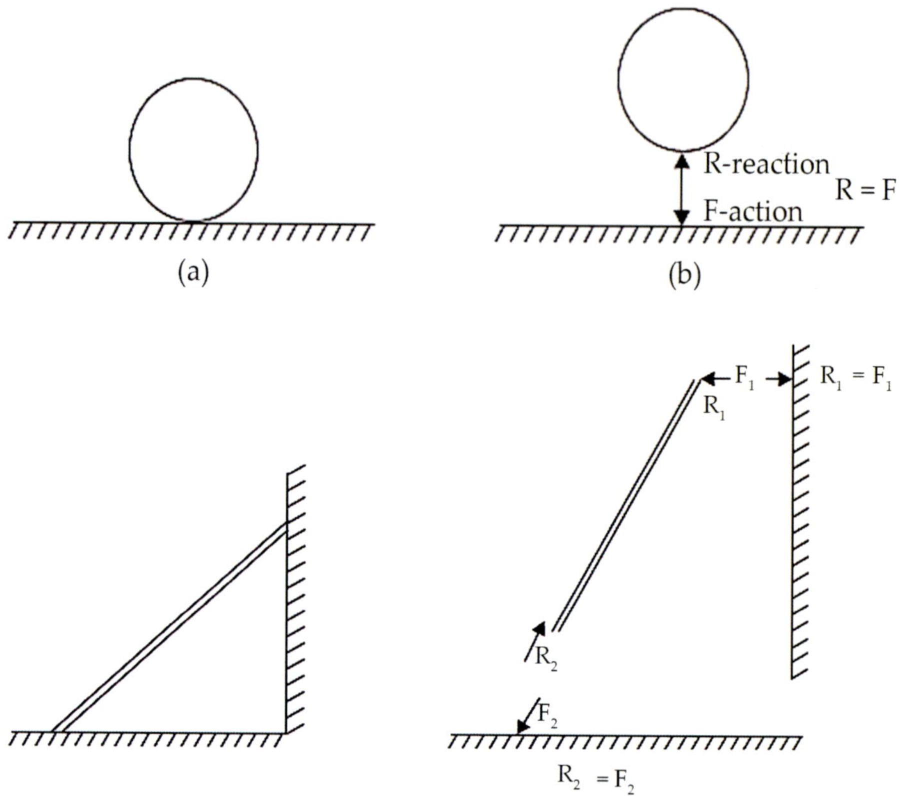

Figure 1.8. Depiction of Newton's second law of motion.

Source: Emertius, Creative Commons License.

1.5.4. Newton's Law of Gravitation

Everybody draws the other body. Any two bodies will attract one another with force directly proportional to their masses and inversely proportional to the square of their distances. The attraction force between the bodies of mass m_1 and mass m_2 at a distance d, is determined by this law.

$$F = G\frac{m_1 m_2}{d^2} \tag{1.4}$$

where, G is the proportionality constant and and is known as the constant of gravitation (Wang, 2007).

1.5.5. Law of Transmissibility of Force

This law states that if a force operating on a rigid body is substituted by another force of the same direction and magnitude acting anywhere along the replaced force's action line, the rigid body's state of rest or motion will remain unchanged.

As shown in Figure 1.9, let F represent the force exerted on a rigid body at point A. The force has the same impact on the state of the body as the force F exerted at point B, according to the law of force transmissibility.

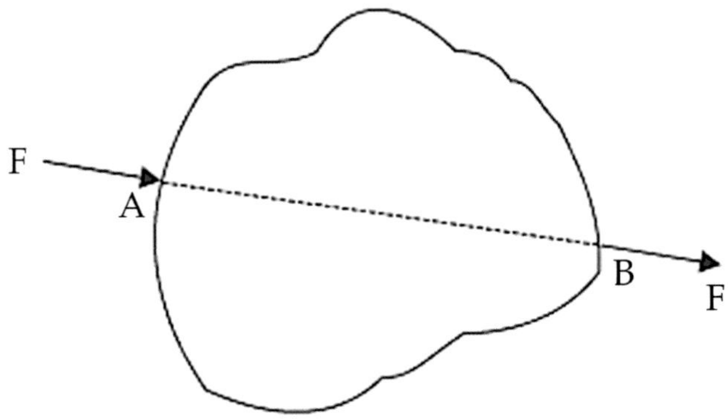

Figure 1.9. Illustration of law of transmissibility of force.

Source: Emertius, Creative Commons License.

While applying the law of transmissibility of forces, it is important to remember that it only applies if the body can be considered rigid. This law is widely utilized since the engineering mechanics in this literature are limited to studying the condition of rigid bodies. The same cannot be said in the field of "solid mechanics," where bodies are viewed as deformable and internal forces are investigated (Zhang et al., 2012).

The law of superposition, which states that the action of one system of forces on a rigid body is unaffected by adding or removing another system of forces in equilibrium, can be used to demonstrate the law of transmissibility of forces (Lage et al., 2014).

Just suppose the rigid body in Figure 1.10(a). It experiences a force of F at A. B is a different location along the force's path of action. The law of superposition makes it clear that the effect of a given force on the body is unaffected if two equal and opposite forces of size F are applied at B along its line of action (*refer* Figure 1.10(b)). A system of forces in equilibrium is created by forces F at A and the opposing force F at B. The system is represented in Figure 1.10(c) if these two forces are removed. We can conclude that the law of force transmissibility is established by studying the system of forces in Figure 1.10(a) and (c) (Law et al., 2011).

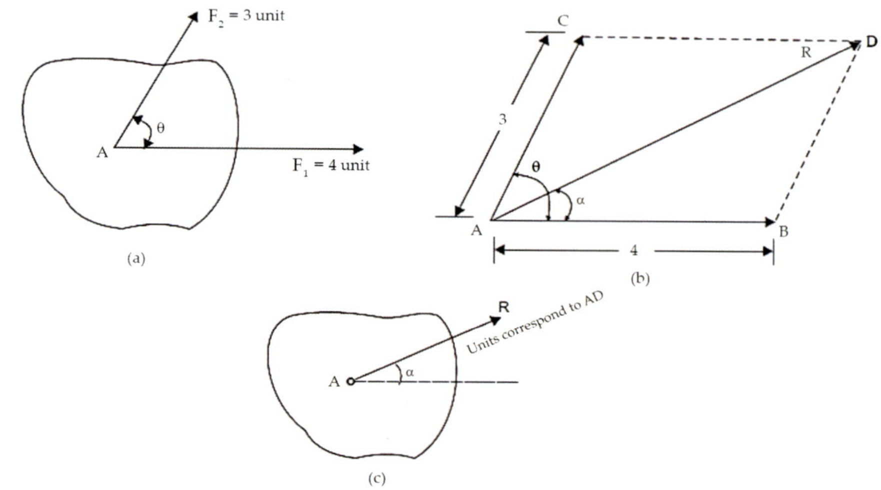

Figure 1.10. Diagram of rigid body with force applying at one end.

Source: Wondrium Daily, Created Commons License.

1.5.6. Parallelogram Law of Forces

The parallelogram law of forces allows us to identify a single force, referred to as the resultant, which can function in place of two other forces at a point and provide the same results. Following the results of experiments, this law was created. Although Stevinces used it in 1586, Varignon & Newton (1687) are credited with presenting it as a law. According to this rule, if two forces exerting simultaneously on a body at a point are signified by the two adjacent sides of a parallelogram in magnitude and direction, their resultant is expressed by the parallelogram's diagonal, which acts through the point of the connection between the two sides representing the forces, in magnitude and direction (Jia et al., 2021).

Forces F_1 = 4 units and F_2 = 3 units are exerted on a body at location A in Figure 1.11. The parallelogram ABCD is then built with AB equal to 4 units on a linear scale and AC equal to 3 units to obtain the consequence of these forces. This law demonstrates that, the diagonal AD indicates the resultant in the direction and magnitude. As a result, the sum of the effects of the forces F_1 and F_2 on the body is equal to units AD in the direction of F_1, where F_1 is the applied force (Saunders, 1964).

1.6. DERIVED LAWS

We may obtain the resulting AD by generating triangle ABD, as shown in Figure 1.11(b). Lines AB and BD, respectively, represent

F_1 and F_2. AD should thus stand for the offspring of F_1 and F_2. After that, we derived the triangle law of forces from the parallelogram law of forces. You can express the Triangle Law of Forces as when the sides of a triangle depict two forces operating on a body, one after the other; their result is represented by the triangle's closing side, measured from the first to the final point (Moyne & Murad, 2002).

The triangle law of forces allows two forces to be integrated simultaneously when more than two forces operate simultaneously on a body. The resultant of all the forces exerted on the body may then be determined.

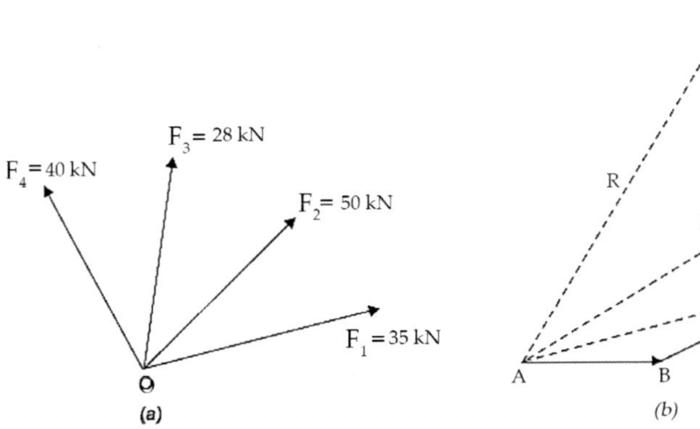

Figure 1.11. Illustration of components of force and resultant force.

Source: Emertius, Creative Commons License.

Figure 1.11 depicts a system of four simultaneous forces applying on a body. AB and F_2 represent F_1 by BC. Hence, following the triangle law of forces, AC refers to the product of F_1 and F_2, or R_1.

If CD is drawn to represent F_3, then AD represents the product of R_1 and F_3 according to the triangle law of forces. In other words, AD is the resultant of F_1, F_2, and F_3. Consider it as R_3 (Wendlandt & Marsden, 1997).

According to the same reasoning, if DE represents F_4, AE represents the product of F_1, F_2, F_3, and F_4. The polygon ABCDE's closing line in the direction of AE thus represents the resultant R. The polygon of the law of forces is what we have derived. It can be stated as follows: "If several simultaneous forces exerting on a body at the same time are signified in magnitude and direction by the sides of a polygon, taken in a particular order, then the resultant is expressed in magnitude and direction by the closing side of the polygon, brought from the first to the last point (Torchigin & Torchigin, 2016)

1.7. UNITS

The fundamental units of mechanics are time (S), mass (M), and length (L). All other units can be drawn from these fundamental units. The three most often employed engineering systems are:

- Meter kilogram second (MKS);
- Centimeter gramme second (CGS); and
- Foot pound second (FPS).

The system's name is based on the employed time, mass, and length units. The units for other variables can be determined using these fundamental units. For instance, the units for the different quantities in MKS are as in Table 1.1 (Heo et al., 2008).

Table 1.1. Tabular Representation of Physical Quantities with Their Units

Quantity	Unit	Notation
Acceleration	Meter per second square	m/sec^2
Velocity	Meter per second	m/sec
Volume	Cubic meter	m^3
Area	Square meter	m^2

Source: Bhavikatti S., Creative Commons License.

1.7.1. Forces Unit

The entire world is transitioning to the SI system of units. SI, sometimes known as the International System of Units, stands for System Internationale d'Unités. Like the MKS system, the fundamental units in the SI system are the second for time, kilogram for mass, and meter for length. The choice of the unit of force is where the MKS and SI systems diverge (Persson, 2012). From Eqn. (1.3), we have

$$\text{Force} \propto \text{Mass} \times \text{Acceleration} = k \times \text{Mass} \times \text{Acceleration} \qquad (1.5)$$

The force that enables a 1 kg mass to flow with an acceleration of 1 m/sec^2 is defined as 1 Newton in the SI system. As a result, the proportionality constant k changes to units. Eqn. (1.5) can be used to draw a unit of force as

$$\text{Unit of Force} = m/sec^2 \times kg = kg - m/sec^2$$

In MKS, a force unit is defined as the force needed to accelerate a mass of 1 kg at the rate of "g" m/sec^2. The name for this unit of force is kilogram weight, or kg/wt. Near the surface of the earth, gravitational acceleration is 9.81 m/sec^2. The change in gravitational acceleration is small and can be regarded as 9.81 m/sec^2 in all engineering mechanics situations. Since 9.81 is the constant of proportionality in Eqn. (1.5), this means:

1 kg-wt = 9.81 newton(1.6)

As a point of interest, kg-wt force is only considered kg in public (Xiao et al., 2009).

1.7.2. Unit of Constant of Gravitation

Experimental findings have demonstrated that G = 6.673 × 10^{-11} Nm^2/kg^2. Hence, if two bodies, one weighing 10 kg and the other 5 kg, are distanced by 1 m, they will exert a force on one another.

From Eqn. (1.4),

$$F = G\frac{m_1 m_2}{d^2} \quad \text{or} \quad G = \frac{Fd^2}{m_1 m_2}$$

$$\text{Unit of } G = \frac{\text{N} \times \text{m}^2}{\text{kg} \times \text{kg}} = \text{Nm}^2/\text{kg}^2$$

According to experimental results, G = 6.673 × 10^{-11} Nm^2/kg^2. As they are distanced by 1 m, two bodies—one weighing 10 kg and the other 5 kg—will exert a force on one another (Quinn & Speake, 2014).

$$F = \frac{6.673 \times 10^{-11} \times 10 \times 5}{1^2} = 33.365 \times 10^{-10} \text{ N}$$

Let's now determine the force operating between a mass of 1 kg near the surface of the earth and the earth. The radius of Earth is 6,371 × 103 meters, and its mass is 5.96506 × 1,024 kg. The force between the two particles is hence,

$$= \frac{6.673 \times 10^{-11} \times 1 \times 5.96504 \times 10^{24}}{(6371 \times 10^3)^2} = 9.80665 \text{ N.}$$

Often, we refer to a body's weight when referring to the force the earth applies to it. Hence, the weight of a kilogram of mass on the planet's surface is 9.80665 newtons or roughly 9.81 newtons in all practical situations. The example of two bodies of 10 kg and 5 kg mass demonstrates how minimal the force applied by two bodies close to the earth's surface is compared to this force. Using the Eqn. (1.4), we can denote the body's weight as W. We get:

$$W = \frac{GmM_e}{r^2}$$

where; *m* denotes the body's mass; M_e denotes the earth; s mass; and *r* denotes the earth's radius.

Denoting $\frac{GM_e}{r^2}$ by g, we get

$$W = mg = 9.81 \text{ m} \quad \text{...(1.7)}$$

Unit of g can be obtained as follows:

$$g = \frac{GM_e}{r^2}$$

$$\text{Unit of } g = \frac{\text{Nm}^2}{(\text{kg})^2} \times \frac{\text{kg}}{\text{m}^2} = \frac{\text{N}}{\text{kg}}$$

as unit of Newton force is kg-m/sec^2, we get

$$\text{Unit of } g = \frac{\text{kgm/sec}^2}{\text{kg}} = \text{m/sec}^2$$

Hence g may be referred to as Gravitational acceleration. This acceleration is experienced by anyone falling freely close to the earth's surface. As given in Eqn. (1.7), the value of g is 9.81 m/sec^2 close to the earth's surface (Finkelstein, 1970). Table 1.2 shows the SI system's prefixes when values are too high or too low.

Table 1.2. Tabular Form of Multiplying Factors and Prefixes in SI

Multiplying Factor	Prefix	Symbol	Multiplying Factor	Prefix	Symbol
10^{12}	tera	T	10^{-3}	milli	m
10^{9}	giga	G	10^{-6}	micro	μ
10^{6}	mega	M	10^{-9}	nano	n
10^{3}	kilo	k	10^{-12}	pico	p
10^{0}	–	–	10^{-15}	femto	f
–	–	–	10^{-18}	atto	a

Source: Bhavikatti S., Creative Commons License.

1.8. SOME COMMON FEATURES OF A FORCE

The force is the organization that attempts to alter the stress or uniform motion of the body, according to Newton's first law. The force necessary to create a unit acceleration in a body of a given mass is how we got at the practical definition of a unit force from Newton's second law of motion. Hence, a force of one newton is needed to accelerate a body of mass one kilogram at a rate of one meter per second. It should be emphasized that a force is only fully stated when the four qualities listed below are characterized (Babajide et al., 1997):

- Magnitude;
- Direction;
- Line of action;
- Point of application.

A ladder is propped up against a wall in Figure 1.10-AB. A person of weight 600 N is standing at position C. The characteristics of the force used by the person on the ladder are as follows:

- 600 N magnitude;
- Point of application is at *C,* with 2 m from *A* along the ladder;
- A vertical line of action with a downward direction.

As you can see, the force's magnitude is printed next to the arrow. The arrow's head symbolizes the point of application and the force's direction, while the arrow's line indicates the line of application (Tordesillas et al., 2010).

1.9. FORCES SYSTEM

A system of forces is created when multiple forces interact with a body at once. The system of forces in space comprises all the forces in a system that do not exist in a single plane. A force system is called a coplanar if all its forces are contained within a single plane.

A system is said to have simultaneous forces when all of its forces intersect at the same point—all of the forces in a system run parallel. A collinear force system is one in which all the forces act along the same single path (Sia et al., 2009).

1.10. DIMENSIONAL HOMOGENEITY

Dimension and unit are terms used to denote the qualitative and quantitative descriptions of physical variables, respectively. Many relationships between the physical quantities are discovered. There could be terms with dimensions or some without dimensions. However, the terms on any equation's sides must be the same size. Dimensional homogeneity refers to this concept. Dimensional analysis is the area of mathematics that deals with the dimensions of quantities (Okamura, 2015).

The gravitational system and absolute system are two-dimensional analysis systems. The fundamental units chosen in an absolute system are mass, length, and time. Thus, it is called the MLT system. The core aspects of a gravitational system are force, length, and time. Thus, it is known as the FLT system.

The dimension of acceleration is:

$$\frac{L}{T^2} = LT^{-2}$$

Since its unit is m/sec^2, we have physical relation from Newton's law.

Force = Mass × Acceleration

Hence the dimensional relation is,

$$F = \frac{ML}{T^2} \quad \text{...(1.8a)}$$

or

$$M = \frac{FT^2}{L} \quad \text{...(1.8b)}$$

To convert dimensions between different measurement systems, use Eqn. (1.8). Table 1.3 provides a list of dimensions of some physical quantities (Cohen et al., 2011).

Table 1.3. Tabular Representation of Dimensions of Some Physical Quantities

Quantity	MLT-System	FLT-System
Velocity	LT^{-1}	LT^{-1}
Acceleration	LT^{-2}	LT^{-2}
Area	L^2	L^2
Momentum	MLT^{-1}	FT
Force	MLT^{-2}	F
Volume	L^3	L^3
Gravitational constant	$M^{-1}L^3T^{-2}$	$F^{-1}L^4T^{-4}$

Source: Emertius, Creative Commons License.

- **Checking Dimensional Homogeneity:** In an equation, all terms on the left and right sides should have the same dimensions, as was previously established. For example, if,

X = Y + Z

the terms *X*, *Y*, and *Z* must have the same dimension. If,

X = bY

and if dimensions are not the same for *X* and *Y*, *b* is not a dimensionless constant. In various unit systems, this constant will have a different value.

Example 1.1: Determine that the equation $v^2 - u^2 = 2as$ has dimensional homogeneity:

Here v denotes the final velocity, u denotes the initial velocity, a denotes acceleration, and s denotes the distance moved.

Solution:

Dimension of distance =L

Dimensions of velocity $=LT^{-1}$

Dimensions of acceleration $=LT^{-2}$

By putting these values in the given equation, we have

$LT^{-1} - LT^{-1} = 2LT^{-2}\,L$

$LT^{-1} \equiv LT^{-1} \equiv LT^{-1}$

Hence it is a dimensionally homogeneous equation (Przybylska, 2007).

Example 1.2: Proof that in the given equation, 9.81 is a dimensionless constant. If it is not so, what must be its dimension?

$$s = ut + \frac{1}{2}\,9.81\,t^2$$

where; 's' denotes the distance; 'u' denotes the initial velocity; 't' denotes the time.

Solution: Dimensions of different values are:

$u = LT^{-1}$

$s = L$

$t = T$

putting values in the given equation, we have,

$$L = LT^{-1}\,T + \frac{1}{2} \times 9.81\,T^2$$

$$L = L + \frac{1}{2} \times 9.81\,T^2$$

Hence, 9.81 must not be a dimensionless constant. Its dimension is written as follows:

$$L \equiv \frac{1}{2}\, 9.81\, T^2$$

9.81 must have the same dimensions as acceleration, LT-2. As expressed in m/sec^2, this is the gravitational acceleration term in SI units. Thus, the presented equation cannot be applied immediately to the FPS or CGS systems (Siano, 1985).

1.11. IDEALIZATIONS IN MECHANICS

When applying the concepts of mechanics to actual concerns, several ideal circumstances are taken for granted. It is impossible to come up with workable solutions without making such assumptions. Engineering mechanics frequently makes the following conceptual frameworks:

- The body is rigid.
- The body is capable of being treated as a continuum.
- The body might be viewed as a particle if its size is modest concerning the other distances at play.
- It is possible to treat a force as a point force if the area across which it is operating on a body is small in comparison to the size of the body. For instance, a guy weighs 600 N, as shown in Figure 1.9. The individual is unable to support himself on a single point. There is a small area of the touch relative to the other dimensions of the problem, yet there is some area of contact. As a result, the man's weight is considered a point load.
- Idealized support situations include simple, hinged, fixed, etc. (Liu, 2004).

TIPS

Using a combination of idealizations and realistic models can give a more complete understanding of a system's behavior.

1.12. CALCULATIONS ACCURACY

As mentioned, various ideal circumstances are assumptions made when using mechanics principles to solve engineering challenges. These assumptions impact the accuracy of the results in some way. Many loads in engineering analysis are estimated loads. Imperfection can be found in the design and manufacture of structures. Engineers multiply the mass by a number (1.5 to 2.25), known as the factor of safety, to account for all these uncertainties. Given all this, it is unnecessary to note calculations with more than four digits. In the computations, it is discovered that 0.2% precision is more than enough (Bauschlicher & Langhoff, 1991).

1.13. APPROACHES TO SOLUTION

Engineering mechanics problems can be solved using either the traditional or vector approaches. The traditional technique makes the issue feel tangible. It gives engineers the confidence to accept the findings offered by others and act fast on the job site (Berbenni et al., 2014). Engineers must develop this kind of problem-solving intuition. Therefore, using a traditional approach to address three-dimensional issues becomes challenging. The vector approach is the best method for analyzing three-dimensional issues. However, the issue's tangible sense is lost, and site engineers' capacity for swift decision-making is not improved. As a result, there are two schools of academics: one supports the traditional approach, and the other the vector approach. The author of this book employed a traditional method to address mechanical engineering issues (Wei, 2001).

ACTIVITY 1.1.

How would you calculate the force required to lift a 500 kg weight using a hydraulic jack that has a piston with a diameter of 10 cm? Assume the system is ideal and neglect any frictional losses.

1.14. SUMMARY

The chapter on Engineering Mechanics covers a variety of topics related to the fundamental principles of mechanics. The chapter begins with an overview of the historical development of mechanics and the importance of the field in various industries. The fundamental terms and laws of mechanics are then introduced, including the Law of Inertia, the Law of Acceleration, and the Law of Action-Reaction. The chapter also explores force systems, dimensional homogeneity, and approaches to solution. The concept of idealization in mechanics is introduced, as well as its limitations. The chapter concludes with a discussion on the importance of accuracy in calculations and the significance of studying elastic behavior in mechanics. Overall, the chapter provides a comprehensive overview of the basic principles and concepts in engineering mechanics and sets the foundation for further studies in the field.

REVIEW QUESTIONS

- What is the difference between kinematics and dynamics in mechanics?
- Explain the concept of equilibrium and give an example.
- What is the Law of Inertia and how does it relate to Newton's First Law?
- How do you calculate the moment of inertia of an object?
- What is the significance of studying elastic behavior in mechanics?
- Describe the difference between a coplanar and a non-coplanar force system.
- What are the limitations of idealization in mechanics?

MULTIPLE CHOICE QUESTIONS

1. What is the study of motion without reference to its causes known as?

a. Kinetics
b. Statics
c. Dynamics
d. Equilibrium

2. Which of the following is NOT a fundamental term in mechanics?

a. Velocity
b. Acceleration
c. Time
d. Force

3. According to Newton's First Law, an object in motion will:

a. continue moving at a constant velocity
b. accelerate in the direction of its motion
c. come to a stop
d. remain at rest

4. Which of the following is NOT a derived law in mechanics?

a. Law of Conservation of Energy
b. Law of Conservation of Momentum
c. Law of Inertia
d. Law of Gravitation

5. The SI unit of force is:

a. kilogram
b. meter
c. newton

d. joule

6. Which of the following is NOT a type of force system?

a. Coplanar force system
b. Concurrent force system
c. Non-coplanar force system
d. Centripetal force system

7. Which of the following is an example of a fluid mechanics problem?

a. Calculating the force required to lift a weight
b. Analyzing the motion of a pendulum
c. Determining the moment of inertia of a rotating object
d. Modeling the flow of water through a pipe

Answers to Multiple Choice Questions

1. (a); 2. (c); 3. (a); 4. (c); 5. (c); 6. (d); 7. (d)

REFERENCES

1. Babajide, A., Hofacker, I. L., Sippl, M. J., & Stadler, P. F., (1997). Neutral networks in protein space: A computational study based on knowledge-based potentials of mean force. *Folding and Design, 2*(5), 261–269.
2. Bauschlicher, Jr. C. W., & Langhoff, S. R., (1991). Quantum mechanical calculations to chemical accuracy. *Science, 254*(5030), 394–398.
3. Bažant, Z. P., Zi, G., & Meyer, C., (2000). Fracture mechanics of ASR in concretes with waste glass particles of different sizes. *Journal of Engineering Mechanics, 126*(3), 226–232.
4. Berbenni, S., Taupin, V., Djaka, K. S., & Fressengeas, C., (2014). A numerical spectral approach for solving elasto-static field dislocation and g-disclination mechanics. *International Journal of Solids and Structures, 51*(23, 24), 4157–4175.
5. Bhatt, C. B., & Rao, N. J., (2005). A framework to generate learning objectives based on Vincenti's categorization of engineering knowledge. *WSEAS Transactions on Advances in Engineering Education, 2*(3), 224.
6. Boys, C. V., (1895). I. on the Newtonian constant of gravitation. *Philosophical Transactions of the Royal Society of London (A)*, (186), 1–72.
7. Burdekin, F. M., & Stone, D. E. W., (1966). The crack opening displacement approach to fracture mechanics in yielding materials. *Journal of Strain Analysis, 1*(2), 145–153.
8. Chen, C. J., & Haik, Y., (2002). Some historical and future aspects of engineering mechanics. *Journal of Engineering Mechanics, 128*(12), 1242–1253.

9. Ching, J., (2009). Equivalence between reliability and factor of safety. *Probabilistic Engineering Mechanics, 24*(2), 159–171.
10. Cohen, D. L., Lo, W., Tsavaris, A., Peng, D., Lipson, H., & Bonassar, L. J., (2011). Increased mixing improves hydrogel homogeneity and quality of three-dimensional printed constructs. *Tissue Engineering Part C: Methods, 17*(2), 239–248.
11. Dubourg, V., Sudret, B., & Deheeger, F., (2013). Metamodel-based importance sampling for structural reliability analysis. *Probabilistic Engineering Mechanics, 33*, 47–57.
12. Durst, F., (2000). Fluid mechanics developments and advancements in the 20th century. In: *Proc. 10th Int. Symp. on Applications of Laser Techniques to Fluid Mechanics, 1*, 1–11.
13. Edmonds, A. R., (1996). *Angular Momentum in Quantum Mechanics* (Vol. 4, pp. 1–23). Princeton university press.
14. Feynman, R. P., (1948). Space-time approach to non-relativistic quantum mechanics. *Reviews of Modern Physics, 20*(2), 367.
15. Finkelstein, A. M., (1970). Variable constant of gravitation and celestial mechanics. *Celestial Mechanics, 2*(2), 237–252.
16. Galili, I., & Tseitlin, M., (2003). Newton's first law: Text, translations, interpretations and physics education. *Science & Education, 12*, 45–73.
17. Haas, C., Waugh, L., & Froese, T., (2007). History and renaissance of construction engineering and management in Canada. *Journal of Construction Engineering and Management, 133*(9), 678–683.
18. Haber, R. B., (1990). Visualization techniques for engineering mechanics. *Computing Systems in Engineering, 1*(1), 37–50.
19. Hassaan, G. A., (2016). Mechanical engineering in ancient Egypt, Part II: Jewelry industry (pectorals). *International Journal of Recent Engineering Science (IJRES), 3*(1), 33–45.
20. Heo, S. M., Haase, E. M., Lesse, A. J., Gill, S. R., & Scannapieco, F. A., (2008). Genetic relationships between respiratory pathogens isolated from dental plaque and bronchoalveolar lavage fluid from patients in the intensive care unit undergoing mechanical ventilation. *Clinical Infectious Diseases, 47*(12), 1562–1570.
21. Jia, G., Huang, H., Wang, S., & Li, B., (2021). Type synthesis of plane-symmetric deployable grasping parallel mechanisms using constraint force parallelogram law. *Mechanism and Machine Theory, 161*, 1–09.
22. Kachanov, L., (1986). *Introduction to Continuum Damage Mechanics* (Vol. 10, pp. 1–13). Springer Science & Business Media.
23. Kirsch, U., (2003). Design-oriented analysis of structures—Unified approach. *Journal of Engineering Mechanics, 129*(3), 264–272.
24. Lage, Y. E., Neves, M. M., Maia, N. M. M., & Tcherniak, D., (2014). Force transmissibility versus displacement transmissibility. *Journal of Sound and Vibration, 333*(22), 5708–5722.

25. Lange, M., (2011). Why do forces add vectorially? A forgotten controversy in the foundations of classical mechanics. *American Journal of Physics, 79*(4), 380–388.
26. Law, S. S., Li, J., & Ding, Y., (2011). Structural response reconstruction with transmissibility concept in frequency domain. *Mechanical Systems and Signal Processing, 25*(3), 952–968.
27. Liu, C., (2004). Approximations, idealizations, and models in statistical mechanics. *Erkenntnis, 60*(2), 235–263.
28. Low, D., & Wilson, K., (2017). Weight, the normal force and Newton's third law: Dislodging a deeply embedded misconception. *Teaching Science, 63*(2), 17–26.
29. Lubarda, V. A., & Hoger, A., (2002). On the mechanics of solids with a growing mass. *International Journal of Solids and Structures, 39*(18), 4627–4664.
30. McCormick, M. E., (2009). *Ocean Engineering Mechanics: With Applications* (Vol. 1, pp. 1–5). Cambridge University Press.
31. Morin, J. B., Gimenez, P., Edouard, P., Arnal, P., Jiménez-Reyes, P., Samozino, P., & Mendiguchia, J., (2015). Sprint acceleration mechanics: The major role of hamstrings in horizontal force production. *Frontiers in Physiology, 6*, 404.
32. Moyne, C., & Murad, M. A., (2002). Electro-chemo-mechanical couplings in swelling clays derived from a micro/macro-homogenization procedure. *International Journal of Solids and Structures, 39*(25), 6159–6190.
33. Muhanna, R. L., Zhang, H., & Mullen, R. L., (2007). Interval finite elements as a basis for generalized models of uncertainty in engineering mechanics. *Reliable Computing, 13*, 173–194.
34. Muvdi, B. B., & McNabb, J. W., (2012). *Engineering Mechanics of Materials* (Vol. 1, pp. 1–17). Springer Science & Business Media.
35. Oakley, D. R., Sues, R. H., & Rhodes, G. S., (1998). Performance optimization of multidisciplinary mechanical systems subject to uncertainties. *Probabilistic Engineering Mechanics, 13*(1), 15–26.
36. Okamura, T., (2015). Effect of material and dimensional homogeneity on thermo-mechanical deformation of brake discs during high-speed braking. *SAE International Journal of Commercial Vehicles, 8*(2015-01-2673), 293–301.
37. Pack, D. J., Avanzato, R., Ahlgren, D. J., & Verner, I. M., (2004). Fire-fighting mobile robotics and interdisciplinary design-comparative perspectives. *IEEE Transactions on Education, 47*(3), 369–376.
38. Pawar, P. Y., & Gawande, S. H., (2012). A comparative study on different types of approaches to text categorization. *International Journal of Machine Learning and Computing, 2*(4), 423.
39. Persson, A., (2012). An unintended side effect of pepper spray: Gender trouble and "repair work" in an armed forces unit. *Men and Masculinities, 15*(2), 132–151.
40. Peters, W. H., Ranson, W. F., Sutton, M. A., Chu, T. C., & Anderson, J., (1983). Application of digital correlation methods to rigid body mechanics. *Optical Engineering,*

22(6), 738–742.

41. Pourciau, B., (2006). Newton's interpretation of Newton's second law. *Archive for History of Exact Sciences, 60*(2), 157–207.
42. Przybylska, M., (2007). Finiteness of integrable n-dimensional homogeneous polynomial potentials. *Physics Letters A, 369*(3), 180–187.
43. Rose, P. L., & Drake, S., (1971). The pseudo-Aristotelian questions of mechanics in renaissance culturel. *Studies in the Renaissance, 18*, 65–104.
44. Saunders, L. R., (1964). Rational deduction in physics: The parallelogram of forces. *The British Journal for the Philosophy of Science, 14*(56), 265–273.
45. Saunders, S., (1998). Time, quantum mechanics, and probability. *Synthese, 114*, 373–404.
46. Seeliger, H. V., (1898). Newton's law of gravitation. *Popular Astronomy, 6*, 40–48.
47. Sia, S., Shibazaki, T., Koga, Y., & Yoshida, N., (2009). Experimental determination of optimal force system required for control of anterior tooth movement in sliding mechanics. *American Journal of Orthodontics and Dentofacial Orthopedics, 135*(1), 36–41.
48. Siano, D. B., (1985). Orientational analysis—A supplement to dimensional analysis—I. *Journal of the Franklin Institute, 320*(6), 267–283.
49. Stein, E., (2013). *The History of Theoretical, Material and Computational Mechanics-Mathematics Meets Mechanics and Engineering* (Vol. 1, pp. 1–16). Springer Science & Business Media.
50. Takagi, H., & Ichihara, Y., (2004). Effect of fiber length on mechanical properties of "green" composites using a starch-based resin and short bamboo fibers. *JSME International Journal Series a Solid Mechanics and Material Engineering, 47*(4), 551–555.
51. Tatum, C. B., (1987). Process of innovation in construction firm. *Journal of Construction Engineering and Management, 113*(4), 648–663.
52. Torchigin, V. P., & Torchigin, A. V., (2016). Physical nature and magnitude of optically induced forces derived from laws of mechanics. *Optik, 127*(15), 5976–5983.
53. Tordesillas, A., O'Sullivan, P., & Walker, D. M., (2010). Evolution of functional connectivity in contact and force chain networks: Feature vectors, k-cores and minimal cycles. *Comptes Rendus Mécanique, 338*(10–11), 556–569.
54. Vu, T. L. A., & Le, T. Q., (2019). Development orientation for higher education training program of mechanical engineering in industrial revolution 4. 0: A perspective in Vietnam. *J. Mech. Eng. Res. Dev, 42*(1), 68–70.
55. Wang, D. J., (2005). Bell chime, dragon washbasin: Modern scientific information hidden in ancient Chinese science and technology. *Technische Mechanik-European Journal of Engineering Mechanics, 25*(1), 9–16.
56. Wei, G., (2001). A new algorithm for solving some mechanical problems. *Computer Methods in Applied Mechanics and Engineering, 190*(15–17), 2017–2030.

57. Wendlandt, J. M., & Marsden, J. E., (1997). Mechanical integrators derived from a discrete variational principle. *Physica D: Nonlinear Phenomena, 106*(3, 4), 223–246.

58. Xiao, S., Stacklies, W., Cetinkaya, M., Markert, B., & Gräter, F., (2009). Mechanical response of silk crystalline units from force-distribution analysis. *Biophysical Journal, 96*(10), 3997–4005.

59. Young, M., & Choice, C., (2007). Maximal velocity sprint mechanics. *Track Coach, 179*, 5723–5729.

60. Yu, M. H., (2002). Advances in strength theories for materials under complex stress state in the 20th century. *Appl. Mech. Rev., 55*(3), 169–218.

61. Zhao, S., Ge, D., Zhao, J., & Xiang, W., (2021). Fingerprint pre-processing and feature engineering to enhance agricultural products categorization. *Future Generation Computer Systems, 125*, 944–948.

CHAPTER 2

DYNAMICS

UNIT INTRODUCTION

In this chapter, we will look at the dynamic of rigid bodies, concentrating on the mathematical theories and laws that control their motion. We will look at rigid body kinematics, the variations between inertial versus non-inertial reference points, and how to calculate the inertia moment. We would also look at rotational motion and torque, including the Euler formula, the finite element technique gyroscopes, or the Runge-Kutta method (Saha et al., 2018).

The research of rigid body dynamics focuses on the movement of things that do not change under the effect of outside forces and preserve their size and shape. The dynamics of rigid bodies define the movement of these things in terms of location, acceleration, and velocity and may be employed to predict a body's trajectory in various conditions (Muhanna et al., 2007).

Inertial reference frames are those in which they do not accelerate, whereas non-inertial frameworks are those in which they can reference accelerate. The variations between such two frames of reference are essential in rigid body dynamics because the physical laws act differently in every instance.

The inertial moment is a fundamental principle in rigid body dynamics utilized to calculate an object's resistance to variations in rotary movement. The inertial moment could be calculated using numerical or integration approaches, depending on the mass distributions and shape of the object (Han et al., 2001). The ability of an object to maintain its rotating motion is described by angular momentum, which is a basic notion in rigid body dynamics. Torque is a concept linked to angular momentum representing the rotating force imparted to a body (Liu & Der Kiureghian, 1986).

An example of a mathematical formula that can be used to explain the rotary movement of a rigid object is known as the Euler equation. This law of angular momentum conservation is the foundation for this method, which enables one to compute the angular acceleration that a body is experiencing. In addition to their significance in rigid body dynamics, gyroscopes are vital in the measurement and upkeep of the spatial configuration of various objects (Pichler & Hellmich, 2011).

In conclusion, we will investigate two numerical approaches frequently applied in rigid body dynamics: the finite element method (FEM) and the Runge-Kutta methodology. Both methods are rather popular. These approaches are utilized to simulate and analyze complicated systems' movement. They can provide helpful knowledge about how stiff things behave under various scenarios (Königsberger et al., 2018) (Figure 2.1).

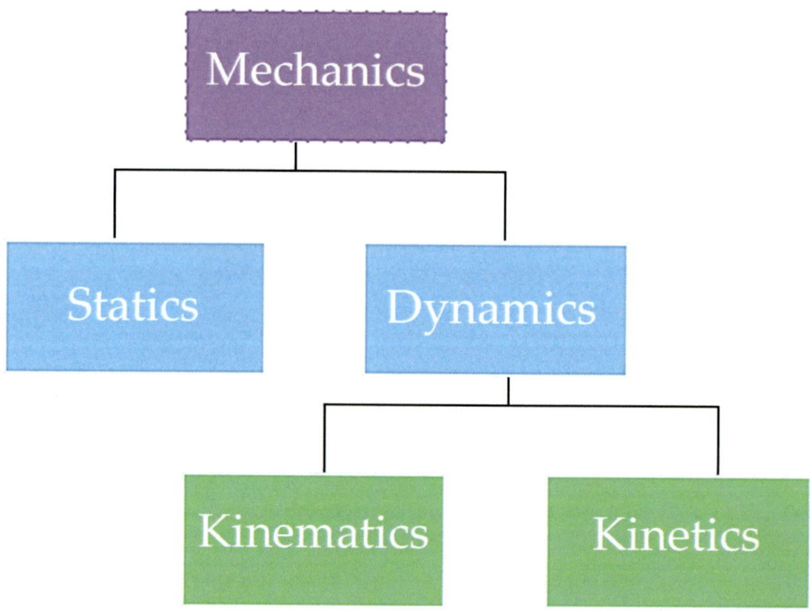

Figure 2.1. Illustration of branches of mechanics.

Source: ME Mechanical, Creative Commons License.

Learning Objectives

At the end of this chapter, readers will be able to:

- Understand the principles of kinematics and its importance in physics.
- Analyze the different aspects of kinematics and their relation to other physical concepts.
- Understand the concept of frame of reference and its significance in describing motion.
- Explain the concept of moment of inertia and its role in rotational motion.
- Analyze the principles of angular momentum and its conservation.
- Understand the Euler equation and its applications in fluid dynamics.
- Analyze the concept of torque-free motion in rigid bodies.
- Understand the principles of Lagrangian approach to rigid body dynamics.
- Analyze the principles of gyroscopes and their applications in physics.
- Understand the numerical methods used to solve problems related to kinematics and dynamics.

Key Terms

- Acceleration
- Angular momentum
- Conservation
- Dynamics
- Euler equation
- Frame of reference
- Gyroscopes
- Kinematics
- Lagrangian approach
- Moment of inertia
- Motion
- Numerical methods
- Rigid body
- Torque-free motion
- Velocity

2.1. KINEMATICS

The field of mechanics known as kinematics is concerned with describing how objects move without considering the forces which move them. In other words, kinematics is focused on investigating the phenomena of motion without considering the causes of movement concerning position, acceleration, velocity, timing, and distance (Bacry & Lévy-Leblond, 1968).

Because it gives a numerical description of movement and enables us to forecast an object's future location and velocity provided by its initial state, kinematics is an essential component of engineering and physics. Kinematics is foundational for more complex subjects like dynamics, examining the forces that generate motion (Robertson, 1935). Rotational and translational kinematics are the two main divisions of kinematics (Robertson, 1936).

2.1.1. Translational Kinematics

The movement of objects along one straight line or in a circular path without any rotation is the subject of translational kinematics. It is focused on the metrics that capture an object's location, acceleration, and velocity. The basic formulas of translational kinematics include (Tsai et al., 1996):

Position: $x = x_0 + v_t + 1/2at^2$

Velocity: $v = v_0 + at$

Acceleration: $a = (v - v_0)/t$

where; 'x' is the location of a body at time t; x_0 is its initial location; 'a' is its acceleration; v_0 is its initial velocity; 'v' is its velocity; and 't' is the time elapsed (Di Gregorio, 2004) (Figure 2.2).

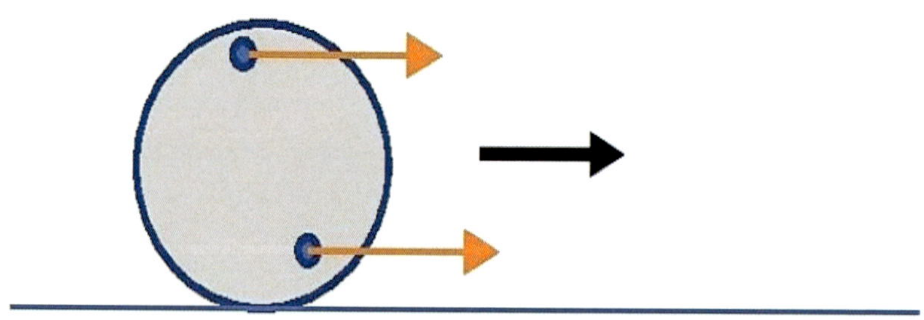

Figure 2.2. Example of translation motion.

Source: Electronic Physics, Creative Commons License.

2.1.2. Rotational Kinematics

On the other hand, rotational kinematics examines the movement of rotational things around an axis. It focuses on the numbers representing the object's angle, angular acceleration, and velocity (Misslisch et al., 1994). The rotary kinematics' basic equations are as follows:

Angular displacement: $\theta = \theta_0 + \omega_t + 1/2\alpha t^2$

Angular velocity: $\omega = \omega_0 + \alpha_t$

Angular acceleration: $\alpha = (\omega - \omega_0)/t$

where; θ is the angular displacement of any object at time t; θ_0 is its initial angular displacement; ω is its angular velocity; α is its angular acceleration; ω_0 is its initial angular velocity; and t is the time elapsed (Tweed & Vilis, 1987).

Kinematics is a basic idea in many disciplines, such as engineering, computer graphics, physics, and robotics. It serves as the basis for the research of dynamics, which focuses on the forces that cause motion and can be utilized to define the movement of anything from subatomic particles to enormous celestial bodies (Bordoloi & Variano, 2017) (Figure 2.3).

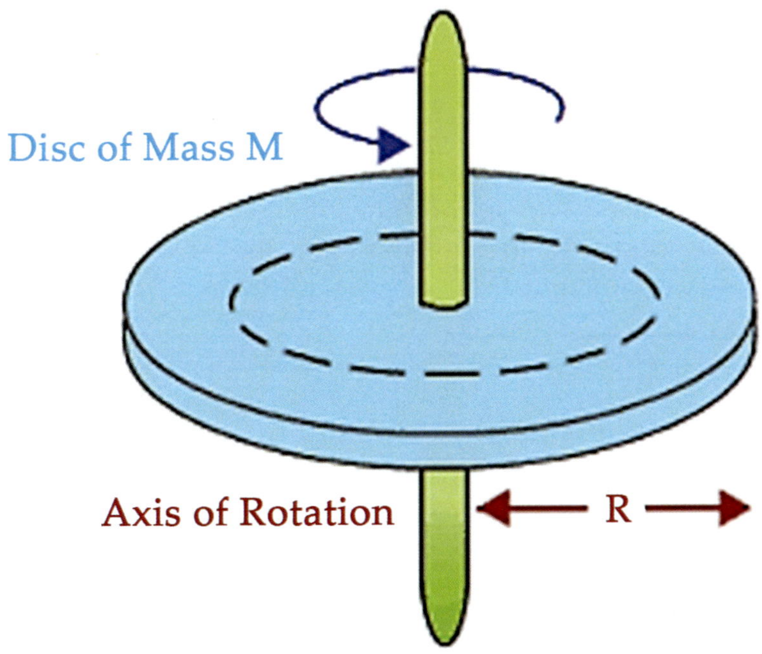

Figure 2.3. Illustration of rotational motion.

Source: Toppr Answr, Creative Commons License.

2.2. ASPECTS OF KINEMATICS

2.2.1. Cartesian Coordinates

Use numeric value; positions in two- or three-dimensional space can be represented with cartesian coordinates. A point is identified in a two-dimensional Cartesian coordinate framework with its location on two perpendicular axes, often called the x-axis and y-axis. An organized pair of values (x,y) is used to represent the location, with "x" standing for the horizontal distance from the point of origin (the location where the x- and y-axes intersect) as well as "y" for the vertical distance to the starting point (Gerdan & Deakin, 1999).

TIPS

Cartesian coordinates can represent not only physical locations but also directions and orientations, which can be useful in applications such as robotics and aerospace engineering. They can also be extended to higher dimensions to represent more complex systems and objects.

Three perpendicular axes, the x-axis, y-axis, and z-axis, are used to find a point in space in a three-dimensional coordinate system called Cartesian. Any distance from the source along the x-axis is represented by "x," while the distance along the y-axis is expressed by "y," and also the distance along the z-axis is expressed by "z." The position is expressed by an arranged triplet of values (x, y, and z) (Salisbury, 1980).

$r = x_i + y_j + z_k$ where i, j, k is fixed orthogonal unit vectors, i.e.:

$i \cdot i = 1, i \cdot j = 0$, etc., e.g., $i = (1,0,0)$.

Don't list the parts while writing about mechanics. r = (x,y,z), but include the basis vectors.

$$r = x_i + y_j + z_k$$

The beginning vectors are occasionally still not constant, and if we only separated the components, the velocity could not be accurate (Civicioglu, 2012).

If $\mathrm{r} = \alpha \mathrm{e}$, then $\dot{\mathrm{r}} = \dot{\alpha}\mathrm{e} + \alpha\dot{\mathrm{e}} \neq \dot{\alpha}\mathrm{e}$ if $\dot{\mathrm{e}} \neq 0$.

eg. $\mathrm{r} = t\mathrm{i} + \mathrm{j} + t^2\mathrm{k}$ with $m = 2$.

$\mathrm{v} = \dot{\mathrm{r}} = \mathrm{i} + 2t\mathrm{k}, \quad v = |\mathrm{v}| = \sqrt{1+4t^2}, \quad \mathrm{p} = m\mathrm{v} = 2\mathrm{v} = 2\mathrm{i} + 4t\mathrm{k},$

$\mathrm{a} = \dot{\mathrm{v}} = 2\mathrm{k}, \quad T = \frac{1}{2}mv^2 = (1+4t^2).$

Note: Acceleration can be non-zero even if the speed is constant since the direction of the velocity might not be constant (Thompson, 1967).

One may get the position by integrating the acceleration, beginning position, and velocity (Figure 2.4). For instance, a = 2k, r(0) = j + k, v(0) = i. v = 2tk + c, but v(0) = c = i therefore v = 2tk + i. r = t^2k + ti + d, but r(0) = d = j + k, therefore r = (t^2 + 1)k + ti + j.

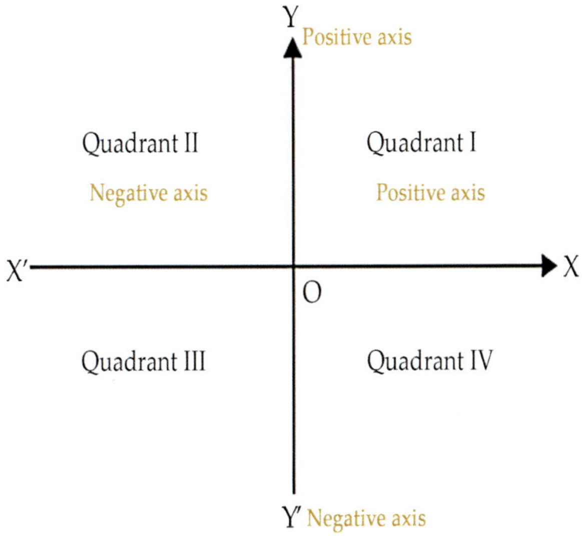

Figure 2.4. Graphical representation of Cartesian coordinate system.

Source: Byjus, Creative Commons License.

2.2.2. Polar Coordinates and Vectors

Consider motion in a plane, using polar coordinates r,θ, where;

x = r cosθ and y = r sinθ. r = xi + yj = r(cosθi + sinθj).

The radial unit vector e_r is a vector in the direction of r,

$e_r = \frac{r}{r} = \cos\theta \mathbf{i} + \sin\theta \mathbf{j}$.

The tangential unit vector $\mathbf{e}_\theta$ is a vector perpendicular to $\mathbf{e}_r$ and $\mathbf{e}_\theta$ = −sinθ**i** + cosθ**j**, (increasing θ is anticlockwise) (Perea et al., 2012).

When a particle moves, r and θ can depend on the time (Nabae & Tanaka, 1996).

$$\dot{e}_r = \frac{d\theta}{dt}\frac{d}{d\theta}e_r = \theta(-\sin\theta i + \cos\theta j) = \theta e_\theta$$

$$\dot{e}_\theta = \frac{d\theta}{dt}\frac{d}{d\theta}e_\theta = \dot{\theta}(-\cos e\theta i - \sin\theta j) = \dot{\theta} e_r$$

$$\text{Note}: e_r \cdot e_r = e_\theta \cdot e_\theta = 1 \text{ and } e_r \cdot e_\theta = 0 \text{ for all time.}$$

$$r = re_r \text{ therefore } \dot{r} = \dot{r}e_r + r\dot{e}_r = \dot{r}e_r + r\dot{\theta}e_\theta$$
$$\ddot{r} = \ddot{r}e_r + \dot{r}\dot{e}_r + \dot{r}\dot{\theta}e_\theta + r\ddot{\theta}e_\theta + r\dot{\theta}\dot{e}_\theta = (\ddot{r} - r\dot{\theta}^2) + (2\dot{r}\dot{\theta} + r\ddot{\theta})e_\theta$$
$$v = \dot{r}e_r + r\dot{\theta}e_\theta, \qquad a = (\ddot{r} - r\dot{\theta}^2) + (2\dot{r}\dot{\theta} + r\ddot{\theta})e_\theta$$

For example. Movement in a circle with continuous speed r = ρ with ρ constant.

$$v = r^{\cdot} = r^{\cdot}e_r + r\theta^{\cdot}e_\theta = \rho\theta^{\cdot}e_\theta.$$

Note that for circular motion v · r = 0 since

$$e_\theta \cdot e_r = 0.\ v = |v| = |\rho\theta^{\cdot}||e_\theta| = |\rho\theta^{\cdot}|$$

Hence constant speed $\theta^{\cdot} = \omega$ with ω constant (choose ω > 0.) v = ωρ hence

$$\omega = v/\rho$$

Since $\theta^{\cdot}$ is constant then $\theta^{\cdot\cdot} = 0$, so $a = -\rho\omega^2 e_r$. Hence

$$a = |a| = \rho\omega^2 = v^2/\rho$$

The acceleration is applied inwards radially. It is called centripetal (or center-seeking) acceleration (Bovolo & Bruzzone, 2006).

Note: Not to be confused with centrifugal (center-fleeing)

2.2.3. Velocity

A key idea in physics called velocity defines how quickly an object's location changes over time. It is also a vector quantity, meaning it contains both a direction and a magnitude (the object's speed) (the direction in which the object is moving) (Loarie et al., 2009).

The formula for velocity is given:

$$v = \Delta x/\Delta t$$

where; "v" is the velocity; "Δx" is the change in location of the object over a particular time interval; and "Δt" is the time interval over which the change in position occurred. The velocity units are usually feet per second (ft/s) or meters per second (m/s), dependent on the system of units used (Billat & Koralsztein, 1996). We could start by defining velocity as the rate at which an object's location changes over time to get the equation for velocity. It may be written numerically as:

$$v = \lim \Delta t \rightarrow 0 \; \Delta x/\Delta t$$

Where limit is determined when Δt approaches zero in time, the instant velocity of an item at a specific time is captured by this definition.

We may use the concept of the derivatives, which refers to the limit of a differential quotient even as the denominator approaches zero, to condense this formula. This results in the following:

$$v = d/dt\ x$$

when; “x” is the position of any object and “d/dt” exposes the derivative about time. This expression provides us with the instantaneous velocity of a body at any given time (Taner & Koehler, 1969).

The following formula can be used to determine an object’s average velocity over a certain period:

$$v_{avg} = (x_f - x_i)/(t_f - t_i)$$

whenever; v_{avg} is the average velocity; x_f is the last position of any object; x_i is the initial position of an object; t_f is the last time; and t_i is the original time. We may find out a body’s current instantaneous velocity using this expression (Wilkie et al., 2009) (Figure 2.5).

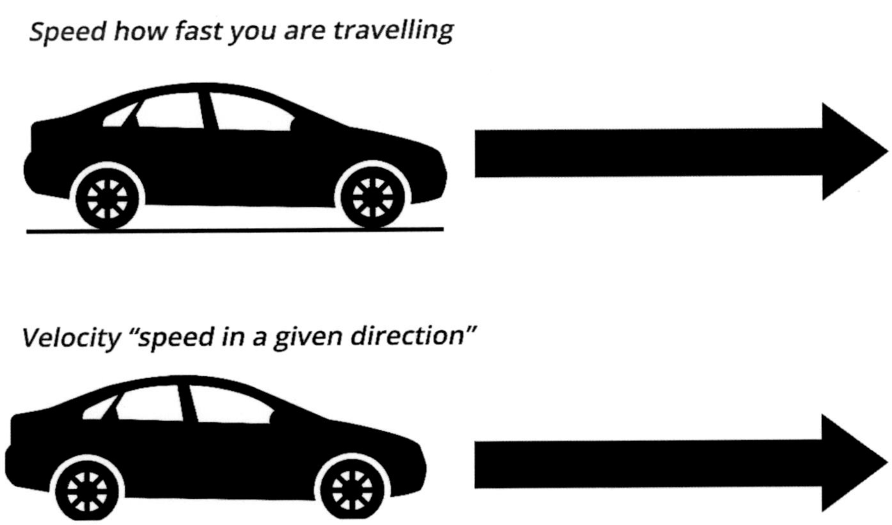

Figure 2.5. Image of velocity.

Source: Cool Kid Facts, Creative Commons License.

2.2.4. Acceleration

Acceleration is a physiological quantity that describes the rate at which the velocity of an item alters over a period. Velocity is a vector quantity, meaning it contains magnitude (the amount of acceleration) and direction (the direction in which the acceleration occurs) (Amendola, 2003).

The rate at which a velocity changes about time is the definition of acceleration in mathematics.

$a = dv/dt$

where; a stands for acceleration; v for velocity; and t for time. According to this formula, acceleration equals the difference between the velocity and time changes (Hill, 1962).

An object's acceleration is positive if its velocity is rising. For instance, an automobile that is moving forward after coming to a stop has positive acceleration. On the reverse side, the acceleration is negative when an object's velocity falls. For instance, a car braking on a stand experiences negative acceleration (Southern & Jones, 2004). In the International System of Units (SI), acceleration is frequently given in units of meters per second squared (m/s^2) (SI). This measurement reflects the velocity changes per unit of time. For instance, the acceleration is 2.5 m/s^2 if an object's velocity rises by 5 m/s in 2 seconds (d'Aspremont et al., 2021).

Other forces operating on an item, including gravity or friction, could also influence acceleration. For instance, gravity dragging something downwards causes a positive acceleration as it is launched into the air. The gravity tugging the ball downward causes it to suffer a negative acceleration as it approaches its highest point and starts to plummet (Rogers & Kimpston, 1992).

2.2.5. Relative Motion (Figure 2.6)

Figure 2.6. Diagram of relative motion.

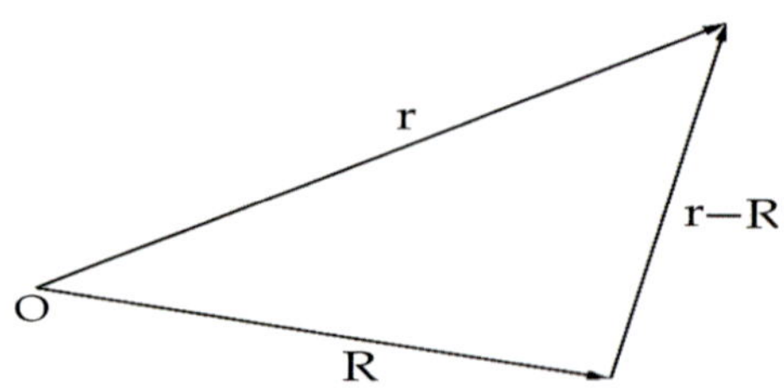

Source: Wojtek Zakrzewski, Creative Commons License.

"r" represents an object's location with a fixed origin. O. Assume an observer is in position R and might be moving. Hence, the object's location of the viewer is (Michelson & Morley, 1887).

$r_e = r - R$

Relative velocity $\dot{r}_e = \dot{r} - \dot{R}$ relative acceleration $\ddot{r}_e = \ddot{r} - \ddot{R}$,

An illustration: With skateboard, Bart travels north (direction j) at 10 mph while experiencing a 25-mph headwind. How fast (velocity) is the wind relative to the ground?

R represents Bart's position, while r represents the position of an air particle (miles and hours are units) (Laube et al., 2005).

$\dot{R} = 10j$, $\dot{r}_e = -25j = \dot{r} - \dot{R}$ $\dot{r} = -25j + 10j = -15j$

The wind is blowing 15 mph southward relative to the earth. What kind of wind does Bart experience as he travels east at 15 mph (Michelson, 1904)?

$$R = 15i, \quad \dot{r} = -15j, \text{so } \dot{\bar{r}} = \dot{r} - R = -15j - 15i$$
$$|\dot{\bar{r}}| = 15\sqrt{2}, \text{so feels a wind of } 15\sqrt{2}\text{mph in the direction} -(i + j)/\sqrt{2} \text{ ie.}$$

2.2.6. Centrifugal Acceleration

That happens when you look at centripetal acceleration with rotational coordinates. Let R represent the position of an observer moving in a circular movement with radius ρ and constant speed v, such as a toddler on a roundabout (Horwitz et al., 1994).

$$\ddot{\mathbf{R}} = -\frac{v^2}{\rho}\mathbf{e}_R$$
hence $$\ddot{\bar{\mathbf{r}}} = \ddot{\mathbf{r}} - \ddot{\mathbf{R}} = \ddot{\mathbf{r}} + \frac{v^2}{\rho}\mathbf{e}_R,$$

So, even for an entity not subject to forces in this plane, $\ddot{r} = 0$, e.g., ball issued by the child, then $\ddot{\bar{\mathbf{r}}} = \frac{v^2}{\rho}\mathbf{e}_R$, As a result, the observer notices a relative acceleration radiating outward. Its acceleration is centrifugal (center-fleeing), e.g., A child notices the ball leaving the ground (Istomin & Gunya, 2020) (Figure 2.7).

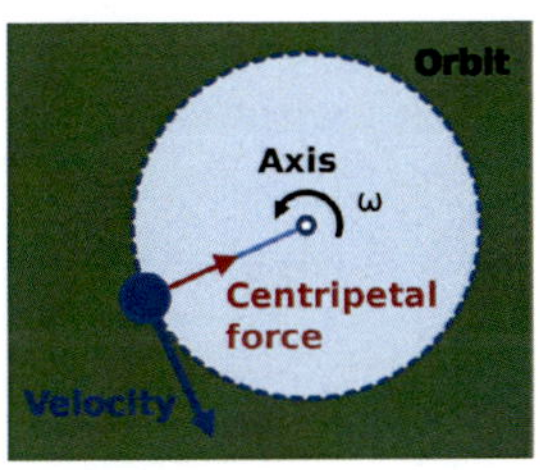

Figure 2.7. Illustration of centripetal force.

Source: Difference Between, Creative Commons License.

2.3. FRAME OF REFERENCE

A reference frame is a coordinate system representing an object's motion in physics. There are two categories of reference frames in classical physics: non-inertial and inertial/rotating frames for reference. Any object not subject to every external force travel in a straight path at a constant speed in an inertial reference frame. In contrast, a non-inertial frame of reference is one where the movement of the frame of reference itself causes an object that is not subject to any external factors to appear to accelerate (Ogbu, 1993).

2.3.1. Inertial Frame

A reference frame at rest or flowing at a constant speed concerning its surroundings in space is called an inertial frame of references. In other terms, an inertial frame of reference would be a frame of reference with which the fundamental physical principles are expressed in their most basic form, i.e., the fundamental physical principles apply to all observers in inertial frames of reference (Alsing et al., 2004).

Remember

A frame of reference is crucial in physics and engineering to accurately describe the motion of an object relative to a specific point or observer. Choosing an appropriate frame of reference can simplify the analysis of the system, and it is important to choose one that accurately represents the object's motion while accounting for any acceleration or rotation.

Classical physics, which defines the movement of things in terms of pressure applied to them, is based on an inertial frame of reference. Unless subject to an outside force, an item will stay still or travel in a straight path with a constant speed by Newton's first law of motion. Only an inertial frame of reference can apply this law (Ciufolini, 2007).

The physical laws, as they are perceived from a frame of reference rotating or accelerating regarding the outside space, would look more complex than the laws as they are viewed from an inertial frame. It is to ensure that additional forces—known as fake forces—that seem to operate on objects inside that non-inertial frame of reference can be introduced by the acceleration or rotations of the frame (Gu et al., 2008). The frame of reference about an item falling freely in a gravitational field illustrates an inertial reference point. The object is at rest regarding the environment inside this frame of reference. The acceleration of the frame causes an imaginary force that is equal to and opposing to the gravitational force that acts on it (Brill & Cohen, 1966).

An object in a spinning system, like a planet or a merry-go-round, has an inertial frame of reference in comparison. The apparent rotational forces, including Coriolis and centrifugal forces, cause extra problems in the motion equations inside this frame of reference. In several branches of physics and engineering, such as classical mechanics**Error! Bookmark not defined.**, relativistic, and astrophysics, an inertial frame of reference is crucial for understanding how physical systems behave (Lauer & Postman, 1994) (Figure 2.8).

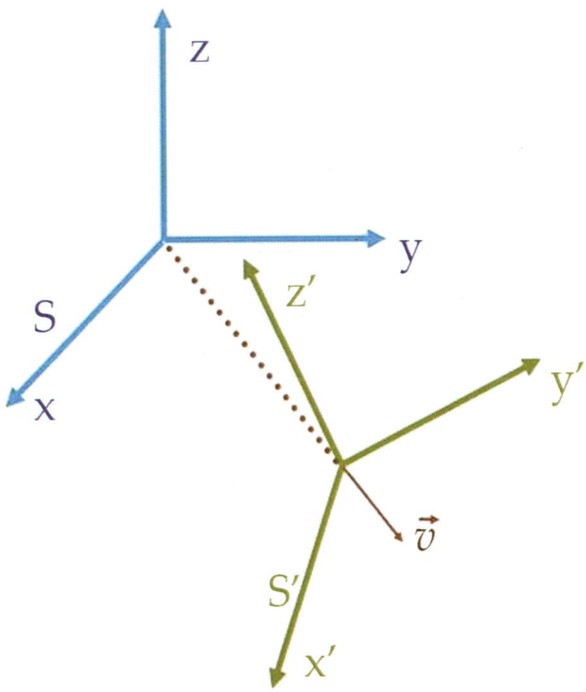

Figure 2.8. Image of inertial frame of reference.

Source: Wiki Audio, Creative Commons License.

2.3.2. Non-Inertial Frames of Reference

In daily life, non-inertial frameworks of reference are frequently encountered. For instance, when an elevator rises or decreases, a man inside feels the same acceleration. As a result, as the elevator accelerates higher or travels lower, the person within will feel heavier and lighter, respectively. In classical physics, a rotational frame of reference, and a non-inertial reference frame, is still a coordinate system that rotates regarding an inertial reference frame. In those other words, in such a stationary observer, the source of the moving frame is traveling on a circular route (Küchemann et al., 2020).

The rules of physics behave differently in a spinning frame of reference than in an inertial reference point. For instance, rotating the frame of reference causes an item that seems to be at rest to accelerate. The imaginary or apparent force, called this acceleration, causes most phenomena seen in rotating systems (Takagi, 1991).

The Coriolis effect is a fundamental idea in studying spinning frames of reference. This apparent deviation of an object's course while observed from such a rotating perspective is known as the Coriolis effect. One reason for this effect is that even though the thing is moving inside a straight line within the rotating frame, the frame is rotating, giving the appearance that such an object's path is curved (Delacroix et al., 2021).

The centrifugal force is a crucial idea in rotational frames of reference. A circularly moving item in a rotational frame of reference experiences what appears to be an outwards force called the centrifugal force. That force develops because of the item speeding toward the rotational center while the frame is rotating, giving the impression that the object is moving away from where it belongs.

Many disciplines, such as fluid dynamics, geophysics, and aeronautical engineering, frequently employ rotational frames of reference. These are especially helpful in cases where an object's motion is strongly related to a system's rotation, including studying rotating equipment or celestial mechanics. Yet, applying rotating frames of reference could also make it more difficult to analyze a system. It may be challenging to reconcile the apparent forces and movement seen in the rotating frame with the real movement of an object in such an inertial reference frame (Vojtech et al., 2012).

2.3.2.1. Comparing Inertial and Rotational System Rates

The impact of a rotational vector basis can be easily considered. Think about the change in the physical quantity that describes a rotating body, db. Two reference frames—one inertial and the other rotating with the body—are used to write this. The rotational variation between these may make a difference (Xu et al., 2019).

$$(db)_{inertial} = (db)_{body} + (db)_{rot}$$

Think about infinitesimal rotational now. We demonstrated the necessity of the conversion matrix's form.

$$O\ (d\theta, \hat{n}) = 1 + d\theta\hat{n} \cdot J$$

where; $[Ji]_{jk} = \varepsilon_{ijk}$

Using this form of J, we may write,

$$[O\ (d\theta, \hat{n})]_{jk} = \delta_{jk} + d\theta n_i \varepsilon_{ijk}$$

We should determine the direction of this rotation. Suppose n^ is in the z-direction, n_i = (0,0,1). Then working on a vector in the xy-plane, say $[i]_i$ = (1,0,0), we have (Gonzalez et al., 2021).

$[O(d\theta,n\hat{})]_{jk}\ i_k \quad = \quad (\delta_{jk} + d\theta n_i \varepsilon_{ijk}) i_k$

$= (i_j + d\theta\varepsilon_{3j1})$

$= (1,0,0) + d\theta\ (0,-1,0)$

Since, ε_{3j1} must have j = 2 to be nonzero, and ε_{321} = −1. The vector turns clockwise after acquiring a negative y-component. Hence, working with results in a rotation that is counterclockwise (positive).

$O\ (d\theta,n\hat{}) = 1 - d\theta n\hat{} \cdot J$

Suppose a vector at time t, b(t) is fixed in a body that rotates along with angular velocity $\omega = \frac{d\theta}{dt}$n. Next, after a time dt, it would have rotated through with an angle dθ = ωdt, so the vector is at time t + dt (Niu et al., 2013).

$b(t + dt) = O(d\theta,n\hat{})\mathbf{b(t)}$

In components,

$b_j\ (t + dt) = (\delta_{jk} - d\theta n_i \varepsilon_{ijk}) b_k\ (t)$

$= \delta_{jk} b_k\ (t) - d\theta ni\varepsilon_{ijk} b_k\ (t)$

$= b_j\ (t) - d\theta n_i \varepsilon_{ijk} b_k\ (t)$

Hence, going back to vector notation,

$= b_j\ (t) - d\theta\ (\varepsilon_{kij} b_k\ (t) n_i)$

$b(t + dt) - b(t) = -d\theta b(t) \times n$

We obtain the change rate by dividing it by dt.

$$\frac{d\mathbf{b}(t)}{dt} = \boldsymbol{\omega} \times \mathbf{b}(t)$$

If b(t) moves concerning the rotating body rather than remaining static inside the rotating system, its change rate is the total of this change and the rate of change brought about by rotation (Johnson et al., 2019).

$$\left(\frac{d\mathbf{b}}{dt}\right)_{inertial} = \left(\frac{d\mathbf{b}}{dt}\right)_{body} + \boldsymbol{\omega} \times \mathbf{b}(t)$$

$$\left(\frac{d}{dt}\right)_{inertial} = \left(\frac{d}{dt}\right)_{body} + \boldsymbol{\omega} \times$$

2.3.2.2. Dynamics in Rotating Frame of Reference

Think of an inertial reference frame and a spinning frame with an origin fixed at the inertial frame's origins. Let r(t) represent a particle's location vector within the rotating reference frame. Therefore, the particle's speed in an inertial reference frame, $\mathbf{v}_{inertial}$, and the frame's rotational speed, $\mathbf{v}_{body}$, are concerned by (Zadravec et al., 2007).

$$\left(\frac{d\mathbf{r}}{dt}\right)_{inertial} = \left(\frac{d\mathbf{r}}{dt}\right)_{body} + \boldsymbol{\omega} \times \mathbf{r}$$

$v_{inertia} = v_{body} + \boldsymbol{\omega} \times \mathbf{r}$

We use the operator once more to determine the acceleration,

$$\frac{d\mathbf{v}_{inertial}}{dt} = \left(\frac{d}{dt} + \boldsymbol{\omega}\times\right)(\mathbf{v}_{body} + \boldsymbol{\omega} \times \mathbf{r})$$

$$= \frac{d(\mathbf{v}_{body} + \boldsymbol{\omega} \times \mathbf{r})}{dt} + \boldsymbol{\omega} \times (\mathbf{v}_{body} + \boldsymbol{\omega} \times \mathbf{r})$$

$$= \left(\frac{d\mathbf{v}}{dt}\right)_{body} + \frac{d\boldsymbol{\omega}}{dt} \times \mathbf{r} + \boldsymbol{\omega} \times \frac{d\mathbf{r}}{dt} + \boldsymbol{\omega} \times \mathbf{v}_{body} + \boldsymbol{\omega} \times (\boldsymbol{\omega} \times \mathbf{r})$$

$$= \left(\frac{d\mathbf{v}}{dt}\right)_{body} + \frac{d\boldsymbol{\omega}}{dt} \times \mathbf{r} + 2\boldsymbol{\omega} \times \mathbf{v}_{body} + \boldsymbol{\omega} \times (\boldsymbol{\omega} \times \mathbf{r})$$

Hence, the accelerations are connected by:

$$\mathbf{a}_{inertial} = \mathbf{a}_{body} + \frac{d\boldsymbol{\omega}}{dt} \times \mathbf{r} + 2\boldsymbol{\omega} \times \mathbf{v}_{body} + \boldsymbol{\omega} \times (\boldsymbol{\omega} \times \mathbf{r})$$

As the inertial frame is consistent with Newton's second law, we get.

$F = ma_{inertial}$

'F' stands for any applied forces. Hence, moving the additional phrases to the left:

$$F = -m\frac{d\boldsymbol{\omega}}{dt} \times \mathbf{r} - 2m\boldsymbol{\omega} \times \mathbf{v}_{body} - m\boldsymbol{\omega} \times (\boldsymbol{\omega} \times \mathbf{r}) = m\mathbf{a}_{body}$$

The Coriolis theorem is this. We give every term thought. The initial,

$$-m\frac{d\boldsymbol{\omega}}{dt}$$

Only apply if the rotational speed changes. It makes perfect sense in that direction if an angular velocity is rising, then $\frac{d\boldsymbol{\omega}}{dt}$ It moves in the rotational direction, and the particle's inertia would prevent this change from occurring. Hence, the direction of the effective fighting force is the reverse (Zhang et al., 2023).

The second term:

$-2m\omega \times v_{body}$

She was also known as the Coriolis force. Remember that the maximum is achieved if the velocity is parallel to the rotating axis. That corresponds to the motion that pushes the particle farther away from the rotating axis for a positive $\mathbf{v}_{body}$ body. The particle would travel too slowly to maintain pace because the velocity needed to remain above a spot on a spinning body grows as the distance from the axis does.

As a result, it appears as though a force is exerting itself in the rotation's anticlockwise direction. Consider an illustration of a particle ascending at the planet's equator. The rising particles will lag because Earth spins from the west to the east, giving the impression that it is moving faster in that direction (Giżynski et al., 2017).

The final term:

$-m\omega \times (\omega \times r)$

The well-known centrifugal force (arising from centripetal acceleration). The path of a centrifugal force is consequently directly in opposition to the rotating axis because, for Earth's rotation $\omega \times r$, r is the direction of such velocity of an object rotating with Earth.

The result is brought on by the body's propensity to travel straight toward the inertial frame and away from the axis. The centrifugal force radiates outward on a particle at the equator, resisting the gravitational force. Hence, the result of the net acceleration caused by gravity with centrifugal acceleration is (Tyler & Mysak, 1995),

$$\begin{aligned} g_{eff} &= g - \omega^2 r \\ &= 9.8 - \left(7.29 \times 10^{-5}\right)^2 \times 6.38 \times 10^6 \\ &= 9.8 - .0339 \\ &= g\,(1 - .035) \end{aligned}$$

So that a 3.5% decrease in gravitational attraction occurs, Earth cannot be a perfect sphere with an equatorial bulge because the impact is absent only at the poles (Palmer & Massi, 2006) (Figure 2.9).

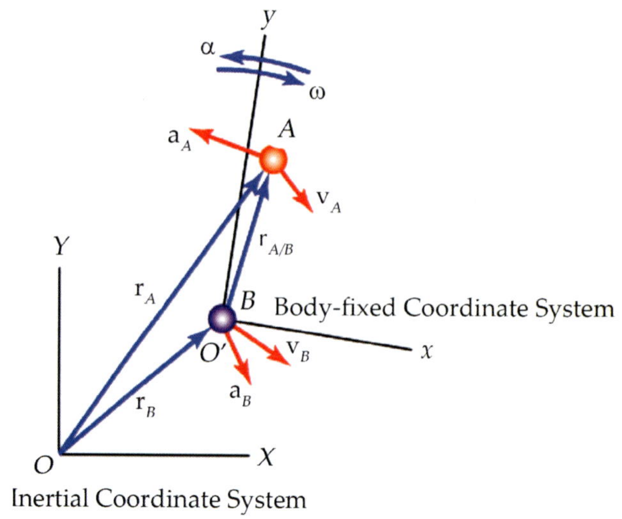

Figure 2.9. Illustration of rotating frame of reference.

Source: Statics, Creative Commons License.

2.4. MOMENT OF INERTIA

Consider a stiff body and fix an arbitrary inertial reference frame. Consider the body's overall torque. Due to internal forces, the ith particle will experience torque (Beliaev, 1961).

N τi = Xri × F_{ji}

j=1

F_{ji} is the force employed by the jth particle on the ith particle. Hence, the double sum represents the overall torque on the body.

$$\begin{aligned} \tau_{internal} &= \sum_{i=1}^{N}\sum_{j=1}^{N} \mathbf{r}_i \times \mathbf{F}_{ji} \\ &= \frac{1}{2}\sum_{i<1}^{N}\sum_{j=1}^{N} (\mathbf{r}_i \times \mathbf{F}_{ji} + \mathbf{r}_j \times \mathbf{F}_{ij}) \\ &= \frac{1}{2}\sum_{i<1}^{N}\sum_{j=1}^{N} (\mathbf{r}_i - \mathbf{r}_j) \times \mathbf{F}_{ji} \end{aligned}$$

The final step uses Newton's third law; in contrast, we assume that such forces between components in the rigid body act down the line connecting the two particles (Bächer et al., 2014).

$$F_{ji} = \frac{r_i - r_j}{|r_i - r_j|}$$

So, all the cross results vanish, and

$\tau_{internal} = 0$

Consequently, when computing torque, we consider external forces operating on the body. Working within the continuum limits is now simpler. To restore the discrete picture for a discrete grouping of masses, let's set the density at every location of the body to be $\rho(\mathbf{r})$. Every external force acting at location r of the body, dF(**r**), contributes to the overall torque (Bischoff & Gross, 2011).

$d\tau = r \times dF(r)$

and by incorporating this, the total is obtained. Using Newton's second law to replace the force, dF(r) = $\frac{d\mathbf{v}}{dt}dm = \frac{d\mathbf{v}}{dt}\rho(\mathbf{r})\,d^3x$, we have:

$$\begin{aligned}\tau &= \int \mathbf{r} \times \frac{d\mathbf{v}}{dt}dm \\ &= \int \rho(\mathbf{r})\left(\mathbf{r} \times \frac{d\mathbf{v}}{dt}\right)d^3x \\ &= \int \rho(\mathbf{r})\left[\frac{d}{dt}(\mathbf{r}\times\mathbf{v}) - \left(\frac{d\mathbf{r}}{dt}\times\mathbf{v}\right)\right]d^3x\end{aligned}$$

Since, $\frac{d\mathbf{r}}{dt} \times \mathbf{v} = \mathbf{v} \times \mathbf{v} = 0$, therefore, the density is not time dependent.

$$\tau = \frac{d}{dt}\int \rho(\mathbf{r})(\mathbf{r}\times\mathbf{v})\,d^3x$$

Because dL is for a tiny mass element, the right-hand side is only the total angular momentum. dm = ρd^3x is dL = ρ(r) (r × v).

Now imagine that the body spins at an angular velocity ω. Then any point's velocity within the body is ω × r, so,

$$\begin{aligned}\tau &= \frac{d}{dt}\int \rho(\mathbf{r})(\mathbf{r}\times(\boldsymbol{\omega}\times\mathbf{r}))\,d^3x \\ &= \frac{d}{dt}\int \rho(\mathbf{r})\left(\boldsymbol{\omega}r^2 - \mathbf{r}(\mathbf{r}\cdot\boldsymbol{\omega})\right)d^3x\end{aligned}$$

We want to distinguish between the rigid body's inherent features and those that depend on its mobility. To accomplish this, we take the integral above and extract ω, the value; however, this needed index notation. Put the equation's elements in writing (Sorensen, 1973).

$$\begin{aligned}\tau_i &= \int \rho(\mathbf{r})\left(\omega_i r^2 - r_i r_j \omega_j\right)d^3x \\ &= \int \rho(\mathbf{r})\left(\omega_j \delta_{ij} r^2 - r_i r_j \omega_j\right)d^3x \\ &= \omega_j \int \rho(\mathbf{r})\left(\delta_{ij} r^2 - r_i r_j\right)d^3x\end{aligned}$$

Look at how the Kronecker delta and the usage of dummies indices enable us to obtain the same index on ω_j in both terms to bring it outside. The instant of inertia tensor is now defined.

$$I_{ij} \equiv \int \rho(\mathbf{r}) \left(\delta_{ij} r^2 - r_i r_j\right) d^3x$$

It is solely dependent on a rigid body. The symmetry of this tensor is.

$I_{ij} = I_{ji}$

Now, the torque calculation may be expressed as

$$\tau_i = \frac{d}{dt}\left(I_{ij}\omega_j\right)$$

As a result, we've demonstrated what angular momentum is.

$Li = I_{ij}\omega_j$

where; the simple equation for movement in an inertial frame:

$\tau = dL/\ dt$

In general, I_{ij} is not equal to the identity, preventing the parallel alignment of angular momentum with angular velocity (Brody, 1985).

2.5. ANGULAR MOMENTUM AND ITS CONSERVATION

A fundamental idea in physics called angular momentum defines how an object rotates. It is described as the result of an object's angular velocity with inertia moment. Angular momentum is denoted mathematically by the letter "L" and therefore is stated in the following formula:

$L = I\omega$

where; "I" stands for inertia moment; and "ω" for angular velocity (Simmons & Burridge, 1981).

The angular momentum conservation theory is a physics principle that asserts that even in the absence of external torques, the overall angular momentum of an isolated system stays consistent. It implies that regardless of every internal movement or interaction, the angular momentum of a system will stay constant if no exterior torques occurred on it (Wolfenstein, 1951).

Knowing the system behavior wherein the rotational motion is significant requires a strong understanding of angular momentum conservation. It is employed to explain various phenomena, such as how spinning tops behave, how heavenly bodies move, and how subatomic particles move (Courtial et al., 1997).

The motion of a spinning top is one of the most prevalent illustrations of the angular momentum conservation concept. If a top is made to spin, it moves along a vertical axis while maintaining a constant angular momentum. The torque caused either by the angular momentum of a top itself balances the torque caused by the gravitational pull that acts on the top.

Planetary motion inside the solar system illustrates how angular momentum is conserved. The lack of any exterior torques causes the angular momentum of planets to stay constant while they revolve around the sun. The planets' angular momentum varies when approaching or departing the sun, but their orbital speed stays constant (Walborn et al., 2004).

Angular momentum preservation is crucial in investigating subatomic particles like protons and electrons. The conserved angular momentum is employed within those systems to describe how particles behave in the earth's magnetic field and how molecules and atoms are structured.

In conclusion, the idea of angular momentum and its conservation is a key tenet of physics that governs the behavior of systems involving rotary movement and the rotational movement of objects. A highly effective tool for comprehending various phenomena, ranging from the behavior of subatomic particles to the movement of spinning tops (Becattini et al., 2008) (Figure 2.10).

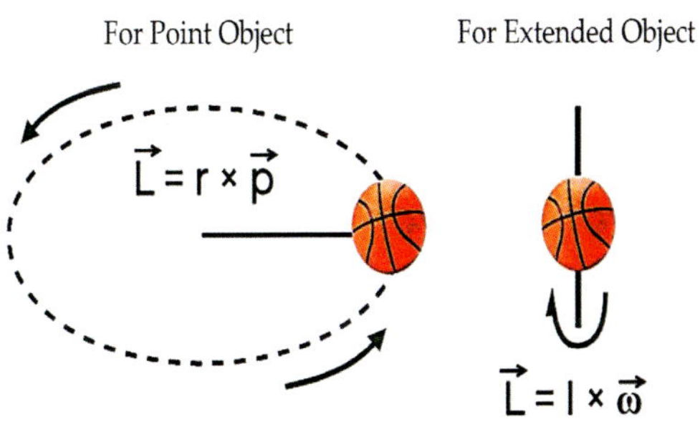

Figure 2.10. Pictorial representation of angular momentum.

Source: Geeksforgeeks, Creative Commons License.

2.6. EULER EQUATION

Consider the motion formula in a reference frame that rotates next. That time derivative must be replaced,

$$\left(\frac{d}{dt}\right)_{inertial} = \left(\frac{d}{dt}\right)_{body} + \boldsymbol{\omega}\times$$

It becomes the motion equation.

$$\tau = \left(\frac{dL}{}dt\right)_b + \omega \times L$$

It is the Euler equation.

Operating inside a specific frame of reference while applying the Euler equation is useful. Every constant orthogonal modification of the basis from our rotating frame leads to a separate rotating frame that is equivalent but has a different basis vector orientation. Additionally, we know that an orthogonal transformation can diagonalize every symmetric matrix. Thus, we might change our basis to one that (Fuhrer & Rudebusch, 2004).

I_{ij} is diagonal. On this basis, we have

$$[I]_{ij} = \begin{pmatrix} I_{11} & 0 & 0 \\ 0 & I_{22} & 0 \\ 0 & 0 & I_{33} \end{pmatrix}$$

The three eigenvalues, I_{11}, I_{22}, and I_{33}, are the primary moments of inertia.

The Euler equation can now be broken down into parts using primary moments (Attanasio & Low, 2004).

$$\tau_i = \frac{d}{dt} I_{ij}\omega_j + \varepsilon_{ijk}\omega_j I_{km}\omega_m$$

writing each part of the sentence separately,

$$\begin{aligned} \tau_1 &= \frac{d}{dt} I_{1j}\omega_j + \varepsilon_{1jk}\omega_j I_{km}\omega_m \\ &= \frac{d}{dt} I_{11}\omega_1 + \varepsilon_{123}\omega_2 I_{3m}\omega_m + \varepsilon_{132}\omega_3 I_{2m}\omega_m \\ &= I_{11}\frac{d\omega_1}{dt} + \varepsilon_{123}\omega_2 I_{33}\omega_3 + \varepsilon_{132}\omega_3 I_{22}\omega_2 \\ &= I_{11}\frac{d\omega_1}{dt} + \omega_2\omega_3\left(I_{33} - I_{22}\right) \end{aligned}$$

and similarly,

$$\begin{aligned} \tau_2 &= \frac{d}{dt} I_{22}\omega_2 + \varepsilon_{231}\omega_3 I_{11}\omega_1 + \varepsilon_{213}\omega_1 I_{33}\omega_3 \\ &= I_{22}\frac{d\omega_2}{dt} + \omega_3\omega_1\left(I_{11} - I_{33}\right) \end{aligned}$$

and

$$\begin{aligned}\tau_3 &= \frac{d}{dt} I_{33}\omega_3 + \varepsilon_{312}\omega_1 I_{22}\omega_2 + \varepsilon_{321}\omega_2 I_{11}\omega_1 \\ &= I_{33}\frac{d\omega_3}{dt} + \omega_1\omega_2\left(I_{22} - I_{11}\right)\end{aligned}$$

Introducing the shorter (yet somewhat deceptive) notation (Giles, 1990).

$I_1 = I_{11}$

$I_2 = I_{22}$

$I_3 = I_{33}$

We have the Euler equations in the form 0

$\tau_1 = I_1\omega^{\cdot}_1 - \omega_2\omega_3 (I_2 - I_3)$

$\tau_2 = I_2\omega^{\cdot}_2 - \omega_3\omega_1 (I_3 - I_1)$

$\tau_3 = I_3\omega^{\cdot}_3 - \omega_1\omega_2 (I_1 - I_2)$

2.7. TORQUE-FREE MOTION

The kinetic energy and angular momentum are maintained whenever the torque is removed. We describe the action to determine the kinetic energy. No such thing as potential; kinetic energy is the integral so over rigid body inside the inertial frame (Kinoshita, 1992).

$$\begin{aligned}T &= \frac{1}{2}\int \rho(\mathbf{r})\,\mathbf{v}^2 d^3x \\ &= \frac{1}{2}\int \rho(\mathbf{r})\,(\boldsymbol{\omega}\times\mathbf{r})\cdot(\boldsymbol{\omega}\times\mathbf{r})\,d^3x \\ &= \frac{1}{2}\int \rho(\mathbf{r})\,(\varepsilon_{imn}\omega_m r_n)(\varepsilon_{ijk}\omega_j r_k)\,d^3x \\ &= \frac{1}{2}\int \rho(\mathbf{r})\,(\varepsilon_{mni}\varepsilon_{jki})\,\omega_m r_n \omega_j r_k d^3x \\ &= \frac{1}{2}\int \rho(\mathbf{r})\,(\delta_{mj}\delta_{nk} - \delta_{mk}\delta_{nj})\,\omega_m r_n \omega_j r_k d^3x \\ &= \frac{1}{2}\omega_m\omega_j \int \rho(\mathbf{r})\,(\delta_{mj} r_n r_n - r_j r_m)\,d^3x \\ &= \frac{1}{2}\omega_m\omega_j I_{mj}\end{aligned}$$

So, we have,

$$T = \frac{1}{2} I_{ij} \omega_i \omega_j$$

The result of the action is $S = \int \frac{1}{2} I_{ij} \omega_i \omega_j dt$

where; $\omega_i = \phi n^{\cdot}{}_i$. Energy since there isn't a clear temporal dependence.

$$\begin{aligned} E &= \frac{\partial L}{\partial \dot{\varphi}} \dot{\varphi} - L \\ &= (I_{ij} n_i n_j \dot{\varphi}) \dot{\varphi} - \frac{1}{2} I_{ij} \omega_i \omega_j \end{aligned}$$

$$= \frac{1}{2} I_{ij} \omega_i \omega_j$$

Being preserved. We also understand the conservation of the total angular momentum,

$$L_i = I_{ij} \omega_j$$

Suppose, for concreteness, that $I_3 < I_2 < I_1$. The simpler scenario of two primary moments being equal would be covered individually. Euler equations then become torque-free motion.

$$I_1 \dot{\omega}_1 = \omega_2 \omega_3 (I_2 - I_3)$$

$$I_2 \dot{\omega}_2 = -\omega_3 \omega_1 (I_1 - I_3)$$

$$I_3 \dot{\omega}_3 = \omega_1 \omega_2 (I_1 - I_2)$$

Where there are non-negative differences, mostly on the right. Multiples of the initial pair are added:

$$I_1 (I_1 - I_3) \omega_1 \dot{\omega}_1 = \omega_1 \omega_2 \omega_3 (I_1 - I_3)(I_2 - I_3)$$

$$I_2 (I_2 - I_3) \omega_2 \dot{\omega}_2 = -\omega_2 \omega_3 \omega_1 (I_1 - I_3)(I_2 - I_3)$$

to find

$$\begin{aligned} 0 &= I_1 (I_1 - I_3) \omega_1 \dot{\omega}_1 + I_2 (I_2 - I_3) \omega_2 \dot{\omega}_2 \\ &= \frac{1}{2} \frac{d}{dt} \left(I_1 (I_1 - I_3) \omega_1^2 + I_2 (I_2 - I_3) \omega_2^2 \right) \end{aligned}$$

So that we have A constant

$$I_1 (I_1 - I_3) \omega_1^2 + I_2 (I_2 - I_3) \omega_2^2 = A$$

Likewise, we discover a connection between ω_3^2 and ω_2^2,

$$I_2 (I_1 - I_2) \omega_2 \dot{\omega}_2 = -\omega_3 \omega_1 \omega_2 (I_1 - I_3)(I_1 - I_2)$$

$I_3 (I_1 - I_3)\omega_3\dot{\omega}_3 = \omega_1\omega_2\omega_3 (I_1 - I_2)(I_1 - I_3)$

So,

$$0 = \frac{1}{2}\frac{d}{dt}\left(I_2 (I_1 - I_2)\omega_2^2 + I_3 (I_1 - I_3)\omega_3^2\right)$$

We determine two elements by solving for the second constant, B.

$$\begin{aligned} I_1\omega_1^2 &= \frac{1}{I_1 - I_3}\left[A - I_2 (I_2 - I_3)\omega_2^2\right] \\ I_3\omega_3^2 &= \frac{1}{I_1 - I_3}\left[B - I_2 (I_1 - I_2)\omega_2^2\right] \end{aligned}$$

Replacing the energy,

$$\begin{aligned} 2E &= I_1\omega_1^2 + I_2\omega_2^2 + I_3\omega_3^2 \\ &= \frac{1}{I_1 - I_3}\left[A - I_2 (I_2 - I_3)\omega_2^2\right] + I_2\omega_2^2 + \frac{1}{I_1 - I_3}\left[B - I_2 (I_1 - I_2)\omega_2^2\right] \\ &= \frac{1}{I_1 - I_3}\left(A + B - I_2 (I_2 - I_3)\omega_2^2 + I_2 (I_1 - I_3)\omega_2^2 - I_2 (I_1 - I_2)\omega_2^2\right) \\ &= \frac{1}{I_1 - I_3}\left(A + B + (-I_2I_2 + I_2I_3 + I_1I_2 - I_2I_3 - I_1I_2 + I_2I_2)\omega_2^2\right) \\ &= \frac{1}{I_1 - I_3}(A + B) \end{aligned}$$

Hence, the energy has a connection to the total of the constants (Lamy & Burns, 1972).

$A + B = 2(I_1 - I_3)E$

To locate the final element, we solve for ω_1 and ω_3,

$$\begin{aligned} \omega_1 &= \sqrt{\frac{1}{I_1 (I_1 - I_3)}\left(A - I_2 (I_2 - I_3)\omega_2^2\right)} \\ \omega_3 &= \sqrt{\frac{1}{I_3 (I_1 - I_3)}\left(B - I_2 (I_1 - I_2)\omega_2^2\right)} \end{aligned}$$

To the differential equation, and substitute for ω_2,

$I_2\dot{\omega}_2 = -\omega_3\omega_1 (I_1 - I_3)$

Once an expression has been integrated, it can be expressed through elliptic integrals.

$$-\sqrt{\frac{I_1I_2I_3}{AB}}t = \int \frac{d\omega_2}{\sqrt{\left(1 - \frac{I_2(I_2-I_3)}{A}\omega_2^2\right)\left(1 - \frac{I_2(I_1-I_2)}{B}\omega_2^2\right)}}$$

Rescale ω_2, letting, then,

$$\chi = \sqrt{\frac{I_2 (I_2 - I_3)}{A}} \omega_2$$

$$-\sqrt{\frac{I_2 (I_2 - I_3)}{}} \sqrt{\frac{I_1 I_2 I_3}{}} t = \int \frac{d\chi}{}$$

where,

$$F(x, k) = \int_0^x \frac{dt}{\sqrt{(1-t^2)(1-k^2t^2)}}$$

$$-\sqrt{\frac{I_2 (I_2 - I_3)}{A}} \sqrt{\frac{I_1 I_2 I_3}{AB}} t = F\left(\sqrt{\frac{I_2 (I_2 - I_3)}{A}} \omega_2, \frac{A (I_1 - I_2)}{B (I_2 - I_3)}\right.$$

We have, and the right side is the first elliptic integral in Jacobi's form.

$$F(x, k) = \int_0^x \frac{dt}{\sqrt{(1-t^2)(1-k^2t^2)}}$$

so we have

$$-\sqrt{\frac{I_2 (I_2 - I_3)}{A}} \sqrt{\frac{I_1 I_2 I_3}{AB}} t = F\left(\sqrt{\frac{I_2 (I_2 - I_3)}{A}} \omega_2, \frac{A (I_1 - I_2)}{B (I_2 - I_3)}\right)$$

Below, we demonstrate how the torque-free solutions are far simpler to comprehend for a symmetrical body (Folgueira & Souchay, 2001).

ACTIVITY 2.1.

What would happen to a spacecraft in a zero-gravity environment if it experiences torque-free motion?

2.8. TORQUE-FREE MOTION OF SYMMETRIC RIGID BODY

Now consider the scenario in which both inertial moments are similar. It occurs whenever the rigid body is symmetrical in rotation about one axis. Let the symmetric axis be the z-axis. Then $I_1 = I_2$, and the (Lamy & Burns, 1972).

Torque-free Euler calculations become:

$$0 = I_1 \dot{\omega}_1 - \omega_2 \omega_3 (I_1 - I_3)$$

$$0 = I_1 \dot{\omega}_2 + \omega_1 \omega_3 (I_1 - I_3)$$

$$0 = I_3 \dot{\omega}_3$$

The last equation demonstrates that ω_3 is fixed. The constant frequency should be defined.

$$\Omega \equiv \omega_3 \left(\frac{I_1 - I_3}{I_1}\right)$$

Two more equations are as follows:

$\omega^{\cdot}_1 = \Omega\omega_2 \; \omega^{\cdot}_2 = -\Omega\omega_1$

We decouple these by subtracting the first and replacing the second (Romano, 2008).

$\omega^{\cdot\cdot}_1 = \Omega^{\cdot}\omega_2$

$= -\Omega^2\omega_1$

likewise, by separating the second and using the first in its place. The pair is the product of this.

$\omega^{\cdot\cdot}_1 + \Omega^2\omega_1 = 0 \; \omega^{\cdot\cdot}_2 + \Omega^2\omega_2 = 0$

with the swift response

$\omega_1 = A\cos\Omega t + B \sin\Omega t$

for ω_1 and, reverting to the initial equation $\omega^{\cdot}_1 = \Omega\omega_2$,

$\omega_2 = -A\sin\Omega t + B \cos\Omega t$

Notice that.

$\omega_1^2 + \omega_2^2 = A^2 + B^2$

Hence, the angular velocity's x and y elements combine to generate a constant-length vector that rotates about the z-axis. If the angular velocity is by ω_3, the remaining elements give the object a "wobble" – it processes as it spins only slightly off its symmetrical axis. In contrast, if ω_3 is small, the movement is a "tumble," a rotation of the symmetrical axis ends over the end. Remember that this study's frame of reference is rotating at angular velocity ω. The object would arrive at resting in the rotating system if all motion occurred about the z-axis. Because our solution for ω is subject to time, even with a frame that rotates with both bodies, bodily functions continue. It spins even if we return to the inertial frame (Adlaj, 2021).

2.9. LAGRANGIAN APPROACH TO RIGID BODIES

We start over and create an action for the issue to analyze symmetric rigid bodies, including one point fixed with gravity operating or tops. We need a set of coordinates to accomplish this. It is assumed that these are the actual Euler angles. A useful collection of three angles can be defined in various ways; we adhere to the one given in Goldstein, (Becker et al., 2009) (Figure 2.11). Relationships between a constant inertial system, (x^0, y^0, z^0) a set of constant Cartesian axes at the top (x,y,z). Three rotational are combined to define the relationship:

1. Rotate at an angle ϕ about the z-axis, giving intermediary coordinates $\xi = (\xi_1,\xi_2,\xi_3)$. We will call that matrix of coordinate transformation D.
2. Turn about the ξ_1 axis by an angle θ, giving matches ξ^0. We will call this matrix of coordinate conversion C.
3. Rotate about ξ_3' by an angle ψ to the final x coordinates. Call this coordinate transformation matrix B.

We could think of (θ,ϕ) as the top's symmetry axis's direction., with ψ stating the direction of its rotation around that axis. Every one of these transformations makes it simple to construct the entire transformation between x^0 and x. D, C, and B are simple 2-dim rotations. Hence, the final transformation, A, is the result (Benson & Hallquist, 1986).

$x. = Ax^0$

$= BCDx^0$

of the three, using D first, then C, then B, were.

	D	=	$\begin{pmatrix} \cos\varphi & \sin\varphi & 0 \\ -\sin\varphi & \cos\varphi & 0 \\ 0 & 0 & 1 \end{pmatrix}$		
	C	=	$\begin{pmatrix} 1 & 0 & 0 \\ 0 & \cos\theta & \sin\theta \\ 0 & -\sin\theta & \cos\theta \end{pmatrix}$		
Multiplying this out, we have A = BCD	B	=	$\begin{pmatrix} \cos\psi & \sin\psi & 0 \\ -\sin\psi & \cos\psi & 0 \\ 0 & 0 & 1 \end{pmatrix}$		
$= \begin{pmatrix} \cos\psi & \sin\psi & & & 0 \\ -\sin\psi & \cos\psi & & & 0 \\ 0 & 0 & 0 & 0 & 1 \end{pmatrix} \begin{pmatrix} 1 & 0 & 0 \\ 0 & \cos\theta & \sin\theta \end{pmatrix} \begin{pmatrix} \cos\varphi & \sin\varphi & & & 0 \\ -\sin\varphi & \cos\varphi & & & 0 \\ 0 & 0 & 0 & 0 & 1 \end{pmatrix}$					
$= \begin{pmatrix} \cos\psi & \sin\psi & & & 0 \\ -\sin\psi & \cos\psi & & & 0 \\ 0 & 0 & 0 & 0 & 1 \end{pmatrix} \begin{pmatrix} \cos\varphi & \sin\varphi & 0 \\ -\cos\theta\sin\varphi & \cos\theta\cos\varphi & \sin\theta \\ \sin\theta\sin\varphi & -\sin\theta\cos\varphi & \cos\theta \end{pmatrix}$					
$= \begin{pmatrix} \cos\psi\cos\varphi - \cos\theta\sin\varphi\sin\psi & \sin\varphi\cos\psi + \cos\theta\cos\varphi\sin\psi & \sin\psi\sin\theta \\ -\sin\psi\cos\varphi - \cos\theta\sin\varphi\cos\psi & -\sin\varphi\sin\psi + \cos\theta\cos\varphi\cos\psi & \cos\psi\sin\theta \\ \sin\theta\sin\varphi & -\sin\theta\cos\varphi & \cos\theta \end{pmatrix}$					

The transposition is all that the inverse transformation is., $A^t = A^{-1}$. Now, indicate the rigid body's angular velocity vector as ω^0. Let the above velocity become the time derivative of a Euler angle with reference towards the inertial reference frame.$(\dot{\varphi}, \dot{\theta}, \dot{\psi})$. By using intermediary coordinates, we could express this relationship as a vector,

$$\omega = \dot{\phi}\hat{z}^0 + \dot{\theta}\hat{\xi}_1 + \dot{\psi}\hat{z}$$

Regarding the body frame, we could locate these parts using A, B, C, and D. For the initial term, $\dot{\phi}\hat{\mathbf{z}}^0$, we write.

$$\hat{\mathbf{z}}' = \begin{pmatrix} 0 \\ 0 \\ 1 \end{pmatrix}$$

and format this for the body of a sentence.

$A\hat{z}^0$	=	$\begin{pmatrix} \sin\psi\sin\theta \\ \cos\psi\sin\theta \\ \cos\theta \end{pmatrix} = \hat{\mathbf{x}}\sin\psi\sin\theta + \hat{\mathbf{y}}\cos\psi\sin\theta + \hat{\mathbf{z}}\cos\theta$
$\dot{\phi}\hat{z}^0$	=	$\hat{x}\dot{\phi}\ \sin\psi\ \sin\theta + \hat{y}\dot{\phi}\ \cos\psi\ \sin\theta + \hat{z}\dot{\phi}\ \cos\theta$

For the next term, $\dot{\theta}\hat{\xi}_1$, to reach the body system, we merely need to complete the final rotation., since $\hat{\xi}_1 = \hat{\xi}_1'$. So, we compute (Glowinski et al., 1998).

$$\begin{aligned} B\dot{\theta}\hat{\boldsymbol{\xi}}_1 &= \dot{\theta}B\hat{\boldsymbol{\xi}}_1' \\ &= \dot{\theta}\begin{pmatrix} \cos\psi \\ -\sin\psi \\ 0 \end{pmatrix} = \hat{\mathbf{x}}\dot{\theta}\cos\psi - \hat{\mathbf{y}}\dot{\theta}\sin\psi \end{aligned}$$

Lastly, $\dot{\psi}\hat{\mathbf{z}}$ is now in the body frame. Adding up these, we have

$$\omega = \dot{\phi}\hat{z}^0 + \dot{\theta}\hat{\xi}_1 + \dot{\psi}\hat{z}$$

$$\begin{aligned} \omega &= \dot{\phi}\hat{z}^0 + \dot{\theta}\hat{\xi}_1 + \dot{\psi}\hat{z} \\ &= \left(\dot{\varphi}\sin\psi\sin\theta + \dot{\theta}\cos\psi\right)\hat{\mathbf{x}} + \left(\dot{\varphi}\cos\psi\sin\theta - \dot{\theta}\sin\psi\right)\hat{\mathbf{y}} + \left(\dot{\psi} + \dot{\varphi}\cos\theta\right)_{\hat{z}} \end{aligned}$$

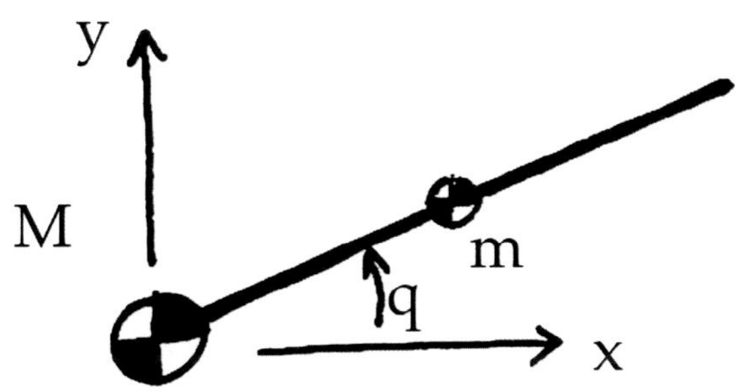

Figure 2.11. Image of Lagrangian formulation.

Source: Stake Exchange, Creative Commons License.

2.9.1. Action Functional for Rigid Body Motion

The Lagrangian and action for one rigid body can now be written. The kinetic energy represents the Euler coordinates (Novikov & Shmel, 1982).

$$T = \frac{1}{2} I_{ij}\omega_i\omega_j$$

When the angular velocity within the major axis frame is substituted, this becomes.

$$T = \frac{1}{2}I_1\left(\dot{\varphi}\sin\psi\sin\theta + \dot{\theta}\cos\psi\right)^2 + \frac{1}{2}I_2\left(\dot{\varphi}\cos\psi\sin\theta - \dot{\theta}\sin\psi\right)^2 + \frac{1}{2}I_3\left(\dot{\psi} + \dot{\varphi}\cos\theta\right)^2$$

We might also possess the center of mass's kinetic energy.

We can express the potential like a function of a mass center alone for a force field that changes slowly; however, torques could also exist if a force field is a gradient or applied at spots mostly on a rigid body. By integrating, one may determine the potential if the object has a gravitational field (Galdi & Silvestre, 2002).

dV = −dmg(r) · dx

where; g(x) is the local gravitational acceleration. For a uniform gravitational field, g = −gk is constant so g(r) · dx = g · x

$$dV = -\left(\rho d^3x\right)\mathbf{g}\cdot\mathrm{x}$$

V = −g · ˆ ρxd³x

= −Mg · R

as the definition of the mass center is:

$$\mathrm{R} = \frac{1}{M}\int \rho\mathbf{x}d^3x$$

We next apply these ideas to the scenario of a rigid body that is symmetrical about one axis and has a fixed position.: tops.

2.9.2. Symmetric Body with Torque: Tops

Now imagine that the rotationally symmetric body is perched, like a spinning top on a tabletop, at one position on the symmetric axis. It serves as a fixed position for us. Gravity at the center of mass produces a torque that acts unless the top is vertical. The potential is when the mass center is l distances well above the fixed tip (Butikov, 2006).

V = Mglcosθ

$I_1 = I_2$ is the result of using the z-axis as the symmetry axis, which means that the initial two terms of the kinetic energy are greatly simplified. The total of the squares combined with the cross terms result in

$$\frac{1}{2}I_1\left(\left(\dot{\varphi}\sin\psi\sin\theta+\dot{\theta}\cos\psi\right)^2+\left(\dot{\varphi}\cos\psi\sin\theta-\dot{\theta}\sin\psi\right)^2\right)=\frac{1}{2}I_1\left(\dot{\varphi}^2\sin^2\theta+\dot{\theta}^2\right)$$

The action then turns into

$$S = \int dt\left(\frac{1}{2}I_1\left(\dot{\varphi}^2\sin^2\theta+\dot{\theta}^2\right)+\frac{1}{2}I_3\left(\dot{\psi}+\dot{\varphi}\cos\theta\right)^2-Mgl\cos\theta\right)$$

We start by searching for conserved quantities. Two angles, ϕ, and ψ, are cyclic; their conjugate momenta are preserved as a result:

$$\begin{aligned} p_\varphi &= \frac{\partial L}{\partial\dot{\varphi}} \\ &= I_1\dot{\varphi}\sin^2\theta+I_3\left(\dot{\psi}+\dot{\varphi}\cos\theta\right)\cos\theta \\ &= \dot{\varphi}\left(I_1\sin^2\theta+I_3\cos^2\theta\right)+I_3\dot{\psi}\cos\theta \end{aligned}$$

and

$$\begin{aligned} p_\psi &= \frac{\partial L}{\partial\dot{\psi}} \\ &= I_3\left(\dot{\psi}+\dot{\varphi}\cos\theta\right) \\ &= I_3\omega_3 \end{aligned}$$

The energy is a third constant of motion since $\frac{\partial L}{\partial t} = 0$. Since the Lagrangian is quadratic in the velocities, we have E = T + V,

$$\begin{aligned} E &= \frac{1}{2}I_1\left(\dot{\varphi}^2\sin^2\theta+\dot{\theta}^2\right)+\frac{1}{2}I_3\left(\dot{\psi}+\dot{\varphi}\cos\theta\right)^2+Mgl\cos\theta \\ &= \frac{1}{2}I_1\left(\dot{\varphi}^2\sin^2\theta+\dot{\theta}^2\right)+\frac{p_\psi^2}{2I_3}+Mgl\cos\theta \end{aligned}$$

To solve, we first remove $\dot{\psi}$,

$$\dot{\psi}=\frac{p_\psi}{I_3}-\dot{\varphi}\cos\theta$$

Then substitute this into p_ϕ,

$$\begin{aligned} p_\varphi &= \dot\varphi\left(I_1\sin^2\theta + I_3\cos^2\theta\right) + I_3\left(\frac{p_\psi}{I_3} - \dot\varphi\cos\theta\right)\cos\theta \\ &= \dot\varphi\left(I_1\sin^2\theta + I_3\cos^2\theta\right) + p_\psi\cos\theta - I_3\dot\varphi\cos^2\theta \\ &= \dot\varphi I_1\sin^2\theta + p_\psi\cos\theta \end{aligned}$$

So that we may also solve for $\dot\phi$, this gives.

$$\dot\varphi = \frac{p_\varphi - p_\psi\cos\theta}{I_1\sin^2\theta}$$

Lastly, we express ourselves through energy. θ as an integral,

$$\begin{aligned} E &= \frac{1}{2}I_1\left(\dot\varphi^2\sin^2\theta + \dot\theta^2\right) + \frac{p_\psi^2}{2I_3} + Mgl\cos\theta \\ &= \frac{1}{2}I_1\left(\left(\frac{p_\varphi - p_\psi\cos\theta}{I_1\sin^2\theta}\right)^2\sin^2\theta + \dot\theta^2\right) + \frac{p_\psi^2}{2I_3} + Mgl\cos\theta \\ &= \frac{1}{2}I_1\dot\theta^2 + \frac{(p_\varphi - p_\psi\cos\theta)^2}{2I_1\sin^2\theta} + \frac{p_\psi^2}{2I_3} + Mgl\cos\theta \end{aligned}$$

The constant term might be dropped., . Then, solving for $\dot\theta$ to integrate,

$$\begin{aligned} t &= \int\frac{d\theta}{\sqrt{\frac{2E}{I_1} - \frac{(p_\varphi - p_\psi\cos\theta)^2}{I_1^2\sin^2\theta} - \frac{2Mgl}{I_1}\cos\theta}} \\ &= \int\frac{\sin\theta d\theta}{\sqrt{\frac{2E}{I_1}\sin^2\theta - \frac{1}{I_1^2}(p_\varphi - p_\psi\cos\theta)^2 - \frac{2Mgl}{I_1}\sin^2\theta\cos\theta}} \end{aligned}$$

or, setting x = cosθ,

$$t = -\int\frac{dx}{\sqrt{\frac{2E}{I_1}(1-x^2) - \frac{1}{I_1^2}(p_\varphi - p_\psi x)^2 - \frac{2Mgl}{I_1}(x - x^3)}}$$

Although this is challenging due to the cubic inside the root, it may be expressed using the concept of elliptic integrals and numerically integrated. The resulting θ (t) then permits us to integrate to get ϕ(t) and ψ (t) (Pina, 2017).

Consider energy a 1-dimensional issue with an effective potential for a more straightforward method of approaching qualitative behavior (Lum et al., 1995).

$$V_{eff} = \frac{(p_\varphi - p_\psi\cos\theta)^2}{2I_1\sin^2\theta} + Mgl\cos\theta$$

Where we once more omit the pointless constant, $\frac{p_\psi^2}{2I_3}$.

To explore the motion in this potential, again set x = cosθ. Then

$$V_{eff} = \frac{(p_\varphi - p_\psi x)^2}{2I_1(1-x^2)} + Mglx$$

It has extrema when.

$$\begin{aligned}
0 &= \frac{dV_{eff}}{dx} \\
&= \frac{-2p_\psi(p_\varphi - p_\psi x)}{2I_1(1-x^2)} - \frac{(p_\varphi - p_\psi x)^2}{2I_1(1-x^2)^2}(-2x) + Mgl \\
&= \frac{1}{2I_1(1-x^2)^2}\left(-p_\psi 2(p_\varphi - p_\psi x)(1-x^2) + 2x(p_\varphi - p_\psi x)^2 + 2I_1 Mgl(1-x^2)^2\right) \\
0 &= -2p_\psi p_\varphi(1-x^2) + 2p_\psi^2 x(1-x^2) + 2p_\varphi^2 x - 4p_\varphi p_\psi x^2 + 2p_\psi^2 x^3 + 2I_1 Mgl(1-2x^2+x^4) \\
0 &= (2I_1 Mgl - 2p_\psi p_\varphi) + (2p_\varphi^2 + 2p_\psi^2)x + (2p_\psi p_\varphi - 4p_\varphi p_\psi - 4I_1 Mgl)x^2 + (2p_\psi^2 - 2p_\psi^2)x^3 + 2I_1 Mglx^4 \\
0 &= (I_1 Mgl - p_\psi p_\varphi) + (p_\varphi^2 + p_\psi^2)x - (p_\varphi p_\psi + 2I_1 Mgl)x^2 + (I_1 Mgl)x^4 \\
0 &= I_1 Mgl(1-x^2)^2 - p_\psi p_\varphi(1+x^2) + (p_\varphi^2 + p_\psi^2)x
\end{aligned}$$

For x near 1, it's possible to ignore the first term, so we have approximately.

$$\begin{aligned}
0 &= p_\psi p_\varphi - (p_\varphi^2 + p_\psi^2)x + p_\psi p_\varphi x^2 \\
x &= \frac{1}{2p_\psi p_\varphi}\left((p_\varphi^2 + p_\psi^2) \pm \sqrt{(p_\varphi^2 + p_\psi^2)^2 - 4p_\psi^2 p_\varphi^2}\right. \\
x &= \frac{1}{2p_\psi p_\varphi}(p_\varphi^2 + p_\psi^2 \pm (p_\varphi^2 - p_\psi^2)) \\
x &= \frac{p_\varphi}{p_\psi}, \frac{p_\psi}{p_\varphi}
\end{aligned}$$

!

In this instance, the top moves at a speed close to its spin rate while processing in a relatively vertical position (Tanriverdi, 2021).

Now think small x, so we ignore the x^4 term. Then

$$\begin{aligned}
0 &= (p_\psi p_\varphi + 2I_1 Mgl)x^2 - (p_\varphi^2 + p_\psi^2)x - (I_1 Mgl - p_\psi p_\varphi) \\
x &= \frac{1}{2(p_\psi p_\varphi + 2I_1 Mgl)}\left((p_\varphi^2 + p_\psi^2) \pm \sqrt{(p_\psi^2 + p_\varphi^2)^2 + 4(p_\psi p_\varphi + 2I_1 Mgl)(I_1 Mgl - p_\psi p_\varphi)}\right. \\
x &= \frac{1}{2(p_\psi p_\varphi + 2I_1 Mgl)}\left((p_\varphi^2 + p_\psi^2) \pm \sqrt{(p_\psi^2 - p_\varphi^2)^2 - 4I_1 Mglp_\psi p_\varphi + 8I_1^2 M^2 g^2 l^2}\right)
\end{aligned}$$

!

It has real solutions as long as

$$(p_\psi^2 - p_\varphi^2)^2 + 8I_1^2 M^2 g^2 l^2 \geq 4I_1 Mglp_\psi p_\varphi$$

While the spin is fast, we have $p_\psi \gg p_\varphi$ and may approximate.

$$p_\psi^4 + 8I_1^2M^2g^2l^2 \geq 4I_1Mglp_\psi^2\frac{p_\varphi}{p_\psi}$$

But then

$$\begin{aligned} 0 &\leq \left(p_\psi^2 - 2I_1Mgl\right)^2 \\ &= p_\psi^4 - 4I_1Mglp_\psi^2 + 4I_1^2M^2g^2l^2 \\ p_\psi^4 + 4I_1^2M^2g^2l^2 &\geq 4I_1Mglp_\psi \end{aligned}$$

And since $\frac{p_\varphi}{p_\psi} < 1$, there is always a solution. The top will proceed in a circle at a set speed whenever the energy and potential equal a certain minimum angle θ. They could consider any changes to this solution. Various solutions of Goldstein, dependent on the relative frequencies in the θ and ϕ oscillations (Naganathan et al., 2014).

Slowly processing top: Think about the situation where we $\dot{\psi} \gg \dot{\theta} \gg \dot{\varphi}$. Then, we could roughly estimate the following:

$$\begin{aligned} \dot{\psi} &= \frac{p_\psi}{I_3} - \dot{\varphi}\cos\theta \\ &\approx \frac{p_\psi}{I_3} \end{aligned}$$

and

$$\dot{\varphi} = \frac{p_\varphi - p_\psi\cos\theta}{I_1\sin^2\theta} \ll \dot{\psi}$$

Whereas the energy is roughly

$$\begin{aligned} E &= \frac{1}{2}I_1\left(\dot{\varphi}^2\sin^2\theta + \dot{\theta}^2\right) + \frac{p_\psi^2}{2I_3} + Mgl\cos \\ E' = \left(E - \frac{p_\psi^2}{2I_3}\right) &\approx \frac{1}{2}I_1\dot{\theta}^2 + Mgl\cos\theta \end{aligned}$$

Then, explaining for θ˙ to integrate, we get an elliptic integral (Wittkowski & Löwen, 2012):

$$\begin{aligned} t &= \int \frac{d\theta}{\sqrt{\frac{2E'}{I_1} - \frac{2Mgl}{I_1}\cos\theta}} \\ &= \sqrt{\frac{2I_1}{E' - Mgl}}F\left(\frac{\theta}{2}; -\frac{4Mgl}{2E' - 2Mgl}\right) \end{aligned}$$

2.10. GYROSCOPES

A force directed perpendicular to a rotational axis act on such a body. A body's orientation changes because of such a force, which causes the rotational axis to move all around the force's axis.

Thinking about the fundamentals of rotational motion can be useful in comprehending gyroscopic motion. Angular momentum results from an object's inertia moment and angular velocity, so it rotates like an axis. Because this angular momentum is maintained, it won't change unless the body is subjected to an external torque (Bergh et al., 1984).

A spinning body experiences a torque whenever an external force is given to it in a perpendicular direction to its rotational axis. The body's angular momentum changes due to this torque, which moves the rotational axis in the opposite direction that the force is exerting (Söderkvist, 1994).

The direction and magnitude of such applied force, the body's inertia moment, and the body's angular velocity are only a few variables that affect the direction or rate of precession. Although it might be complicated to study, body motion is a significant phenomenon in several engineering domains.

Various gyroscopic motion applications range from sports equipment to navigation systems. Engineers could develop better systems for only various applications by knowing the gyroscopic movement principles (Tong & Granat, 1999) (Figure 2.12).

Did you Know?

Gyroscopes use the principles of angular momentum to maintain a stable orientation, making them useful in applications such as navigation and spacecraft stabilization. They are also useful in stabilizing devices that require stable positioning, such as cameras and drones. Understanding gyroscopes is important in physics and engineering.

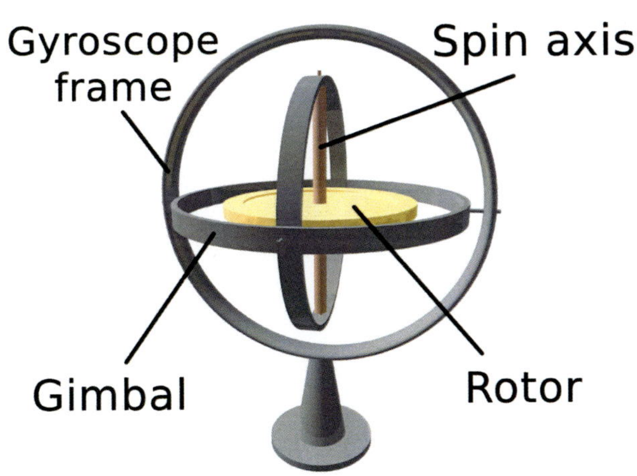

Figure 2.12. Illustration of gyroscope.

Source: Lucas Vieira, Creative Commons License.

2.10.1. Gyroscopic Precession and Nutation

Whenever a gyroscope is exposed to an external force, such as torque, two related processes occur nutation and gyroscopic precession; knowing about the actions of gyroscopes and how they are used in different engineering domains requires understanding such phenomena (Fang et al., 2013).

Whenever a gyroscope is exposed to a force perpendicular to its rotational axis, it moves in a manner known as gyroscopic precession. Stress exerted on a gyroscope causes the rotational axis to process and rotate about a different axis rather than moving the gyroscope in the direction in which the force is applied. It is due to a gyroscope's conserving angular momentum that results in precession when a force is applied since it creates a torque upon that axis of rotation (Han et al., 2020).

The direction and magnitude of the applied force and the gyroscope's moment of inertia determine the precession's direction and pace. Gyroscopic precession could control and stabilize the movement of numerous systems, including airplanes and spacecraft, within engineering applications.

Contrarily, nutation describes the swaying motion that occurs whenever a gyroscope is exposed to an external torque that is not parallel toward its axis of rotation. That motion, which resembles the wobbling of such a spinning top, is brought on by the gyroscope's rotation and the precession of its rotational axis (Armstrong, 1967).

Given that nutation includes the rotations of the overall network and the precession of a gyroscope, it may prove complicated and challenging to study. Nutation must be understood to simulate and manage gyroscope systems, including spacecraft and navigation systems (Böttcher & Henk, 2012).

Here is a straightforward equation derivation to distinguish between nutation and precession: Normally, gyroscopes are installed on freely rotating frames to eliminate external torque. Inside this situation, the potential disappears, and a simpler mechanism is used.

with actual potential

The extrema at,

$$V_{eff} = \frac{(p_\varphi - p_\psi \cos\theta)^2}{2I_1 \sin^2\theta}$$

$$x = \min\left(\frac{p_\varphi}{p_\psi}, \frac{p_\psi}{p_\varphi}\right)$$

where; the value that offers the minimal selection is x ≤ 1. Hence, there is only one viable option.

$$t = \int \frac{d\theta}{\sqrt{\frac{2E}{I_1} - \frac{(p_\varphi - p_\psi \cos\theta)^2}{I_1^2 \sin^2\theta}}}$$

$$= \int \frac{\sin\theta d\theta}{\sqrt{\frac{2E}{I_1}\sin^2\theta - \frac{1}{I_1^2}(p_\varphi - p_\psi\cos\theta)^2}}$$

$$= \int \frac{I_1 dx}{\sqrt{(2I_1E - p_\varphi^2) + 2p_\varphi p_\psi x - \left(2I_1E + p_\psi^2\right)x^2}}$$

As the root is quadratic next time, we could finish the square.

$$2I_1E - p_\varphi^2) + 2p_\varphi p_\psi x - (2I_1E + p_\psi^2)x^2 = -\left(\sqrt{2I_1E + p_\psi^2}x - \frac{p_\varphi p_\psi}{\sqrt{2I_1E + p_\psi^2}}\right)^2 + \frac{p_\varphi^2 p_\psi^2}{2I_1E + p_\psi^2} + 2I_1E - p_\varphi^2$$

Setting

$$\xi = \sqrt{2I_1E + p_\psi^2}x - \frac{p_\varphi p_\psi}{\sqrt{2I_1E + p_\psi^2}}$$

$$dx = \frac{d\xi}{\sqrt{2I_1E + p_\psi^2}}$$

$$A^2 = \frac{p_\varphi^2 p_\psi^2}{2I_1E + p_\psi^2} + 2I_1E - p_\varphi^2$$

$$\Omega = \frac{1}{I_1}\sqrt{2I_1E + p_\psi^2}$$

we have:

$$t = \frac{I_1}{\sqrt{2I_1E + p_\psi^2}} \int \frac{d\xi}{\sqrt{A^2 - \xi^2}}$$

so, we set

and the integral is.

$$t = \frac{I_1}{\sqrt{2I_1E + p_\psi^2}} \sin^{-1}\frac{\xi}{A}$$

$$\sqrt{2I_1E + p_\psi^2}\cos\theta - \frac{p_\varphi p_\psi}{\sqrt{2I_1E + p_\psi^2}} = A\sin\Omega t$$

$$\cos\theta = \frac{p_\varphi p_\psi}{2I_1E + p_\psi^2} + \frac{A}{\sqrt{2I_1E + p_\psi^2}}\sin\Omega t$$

$$= \cos\theta_0 + b\sin\Omega t$$

It illustrates nutation: the gyroscope's tip angle oscillates upwards and downwards all-around angle θ_0 with period Ω. From $\Omega = \frac{1}{I_1}\sqrt{2I_1E + p_\psi^2}$ we see that Ω Depending on what it is of, may be of any magnitude.

I_1. Then, from

$$\dot{\varphi} = \frac{p_\varphi - p_\psi \cos\theta}{I_1 \sin^2\theta}$$

As we can see, the value of p_ϕ makes $\phi^{\cdot}$ independent of Ω. As a result, there is no relationship between the precession rates with nutation (Fadeev et al., 2021).

2.10.2. Applications of Gyroscopic Motion

Gyroscopic motion is a phenomenon that happens if a force is placed parallel to the rotational axis of a rigid body, including a rotational axis. It causes the rotational axis to process around the direction in the force is applied and has several engineering uses. Instances of gyroscopic motion in engineering include the following (Ge & Yu, 2000).

2.10.2.1. Gyroscopes in Navigation Systems

Navigation systems frequently use gyroscopes to assess an object's orientation within space. Each object's positioning and trajectory could be ascertained by detecting variations in orientation with angular velocity using the gyroscopic movement of a rotating rotor. Such technology is employed in aircraft, boats, and spacecraft for smooth flying and precise navigation (Tan et al., 2001) (Figure 2.13).

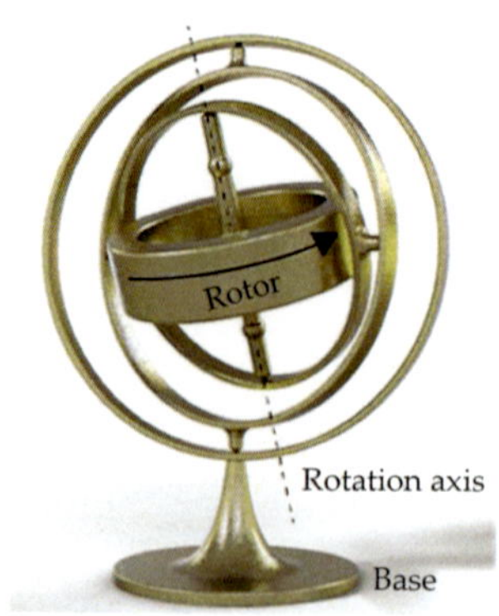

Figure 2.13. Picture of gyroscope in navigation system.

Source: OXTS, Creative Commons License.

2.10.2.2. Gyroscopic Stabilization

Rotating objects around with an axis could be stabilized using gyroscopic motion. Several engineering applications utilize this, including creating centrifuges, washing machines, and turbine engines. A gyroscopic rotor could stabilize and improve any system's motion (Yetkin et al., 2014).

2.10.2.3. Gyroscopic Motion in Robotics

The orientation and movement of robot arms or other mechanical components are measured using gyroscopes within the field of robotics. A robot's motion and location could be controlled using this information, improving its precision and accuracy (Carpenter & Peck, 2009).

2.10.2.4. Gyroscopic Motion in Space Exploration

Gyroscopic motion is essential for space exploration because satellites and spacecraft must maintain an even orientation to operate properly. Gyroscopes are employed to precisely change the spacecraft's orientation with angular velocity to preserve the necessary orientation (Duda & Newman, 2013).

2.10.2.5. Gyroscopic Motion in Sports

Gyroscopic movement is also included in sporting equipment, including tennis racquets and golf clubs. That movement of the object could be stabilized and improved upon by attaching a gyroscopic weight to a grip of a club or racket, leading to more accurate shots and swings (Wagner, 2018).

Gyroscopic motion has numerous engineering uses, ranging from sports goods to navigation. Engineers may develop more efficient devices for various applications by knowing the basic principles of gyroscopic movement and how to use them (Delgado et al., 2021).

2.11. NUMERICAL METHODS FOR SOLVING PROBLEMS

Rigid body dynamics issues are solved using numerical approaches if analytical solutions are unavailable or difficult. These techniques entail employing numerical algorithms that briefly simulate a rigid body's movement to approximate the solution to a motion equation. Problems involving rigid body dynamics can be solved numerically using techniques such as Euler's, Runge-Kutta, and the finite element approach (Kadalbajoo & Gupta, 2010).

1. **Euler's Method:** The basic numerical approach for solving rigid body dynamics issues is Euler's method. By advancing this system's condition

by a small amount of time, it approaches the solution toward the equations of motion. Using Euler's approach, the rigid body's current location and velocity are employed to determine its position and speed at the following time step. Unfortunately, particularly over extended time increments, this strategy could be unstable or may collect errors over time (Chorin, 1997).

2. **Runge-Kutta Method:** For a more precise numerical solution to rigid body dynamics issues, employ the Runge-Kutta technique. It refers to a group of numerical techniques determining the rigid body's speed and position at the following time step by evaluating the equations of motion numerous times at intermediary time steps. This fourth order Runge-Kutta approach, which would be accurate and stable for most rigid body dynamics issues, is the most frequently employed Runge-Kutta variant (Biral et al., 2016).
3. **Finite Element Method (FEM):** The finite element approach would be a numerical technique for solving complex rigid body dynamics issues with variable geometry and boundary conditions. It entails breaking the rigid body into smaller components and roughly estimating the solution of the equation of motion for each component (Davey, 1973). The movement of the rigid body is then ascertained by numerically solving the resultant system of equations. Using the finite element technique is extremely helpful for the design and analysis of structures in which the geometry and loading conditions might be complex (Braun & Sambridge, 1995).

2.11.1. Runge Kutta Method

The fourth level Runge-Kutta technique is a numerical approach to resolving differential equations, such as the movement equations for rigid body dynamics issues. Since it is precise, reliable, and relatively easy to use, it is a well-known and commonly utilized procedure (Butcher, 1996).

The basic idea of the Runge-Kutta method is to use multiple evaluations of the equations of motion at intermediate time steps to calculate the position and velocity of the rigid body at the next step. The fourth order Runge-Kutta method involves the following steps (Butcher, 1976):

Step 1: That acceleration at the present point in time should be calculated:

$a_1 = F/m$

$\alpha_1 = T/I$

If m is the stiff body's mass, F is the total force exerted on it., I it's moment of inertia, T is the net torque acting on it, and a_1 and alpha1 are the linear and angular accelerations, respectively (Anastassi & Simos, 2005).

Step 2: Use the present acceleration to assess the system's status at the halfway point:

$v_1/2 = v(t) + (h/2) * a_1$

$w_1/2 = w(t) + (h/2) * \alpha_1$

where; $v_1/2$ and $w_1/2$ are the respective half-time step linear and angular velocities, and h is the time step size (Zennaro, 1986).

Step 3: Do the half-time step acceleration calculation:

$a_2 = F(r(t) + (h/2) * v(t) + (h^2/4) * a_1)/m$

$\alpha_2 = T(w(t) + (h/2) * w(t) + (h^2/4) * \alpha_1)/I$

where; r is the rigid body's position vector, and a_2 and alpha2 are the linear and angular accelerations at the half-time step, respectively (Hairer, 1978).

Step 4: To use the half-time acceleration, assess the system's condition at the halfway point:

$v_1 = v(t) + h * a_2$

$w_1 = w(t) + h * \alpha_2$

where; v_1 and w_1 are the final values of the time step's angular and linear velocities? respectively.

Step 5: Determine the acceleration so at time step's end:

$a_3 = F(r(t) + h * v_1/2 + (h^2/4) * a_2)/m$

$\alpha_3 = T(w(t) + h * w_1/2 + (h^2/4) * \alpha_2)/I$

where; a_3 and alpha3 after the time step, both the angular and linear acceleration, respectively.

Step 6: Utilizing the end-time acceleration, assess the system's condition after the time step:

$v(t+h) = v(t) + (h/6) * (a_1 + 2a_2 + 2a_3 + a_4)$

$w(t+h) = w(t) + (h/6) * (\alpha_1 + 2\ \alpha_2 + 2\ \alpha_3 + \alpha_4)$

where; a_4 and alpha4 are the angular and linear velocities at the following time step, and v(t+h) and w(t+h) are the angular and linear accelerations determined by evaluating the motion equations just after the time step, respectively (Kraaijevanger, 1991).

Generally, the 4th-order Runge-Kutta technique is a well-liked option for analysis and simulation since it offers relatively stable and accurate numerical results to issues involving rigid body dynamics (Cooper, 1987).

2.11.2. Finite Element Method (FEM)

The FEM is a numerical method to resolve difficult engineering issues involving partial differential equations (PDEs). It is frequently employed in structural, civil, mechanical, and aeronautical engineering. The following steps are involved in FEM (Zhu & Zienkiewicz, 1988):

Step 1: Discretization: The complicated geometry must first be divided into smaller, easier, and finite parts. These components can have a triangular, quadrilateral, tetrahedral, or hexahedral shape. These original PDEs are changed throughout the discretization process into a set of algebraic equations which can be resolved numerically. Nodes, or the places where the elements converge, represent where geometry is divided. Each node is assigned a degree of freedom (DOF), such as displacement or temperature, that is used to describe the behavior of the element (Jagota et al., 2013).

Step 2: Element Equations: The component equations for every finite element must then be derived. The applied loads that nodal displacements are related to an element's DOFs by the element equations. Several techniques can obtain the element equations, including the Galerkin or virtual labor. All element equations that follow from this are represented (Logg, 2007):

$$[K]\{u\}=\{f\}$$

where; [K] is the stiffness matrix; {u} is the nodal displacement vector; and {f} is the nodal force vector.

Step 3: Assembling the System of Equations: The element equations are put together into a mathematical framework for the complete construction in the third stage. That is accomplished by joining the nodes of nearby elements and ensuring that the DOFs are continuous at the shared nodes. This resulting set of equations has the following representation (Babuška, 1973):

$$[K]\{u\}=\{f\}$$

where; [K] is the world stiffness matrix; {u} is the nodal displacement vector globally; and {f} is the vector of the world's nodes (Babuska et al., 1981).

Step 4: Solving the System of Equations: The equation system must be solved to arrive at the nodal displacements with related element strains and stresses inside the final phase. Many numerical techniques, including direct, iterative, and eigenvalue methods, can help solve the equation system. These element equations could be applied to a nodal displacement to compute the strains and stresses (Babuška & Suri, 1990).

Many engineering issues, including heat exchange, fluid flow, electromagnetics, and both dynamic and static structural analysis, could be resolved using the FEM. The size and form of the elements, the arrangement of the component shape functions, and the numerical technique used to solve the equations system all affect how accurate the answer is (Fish, 1992).

Below are some important formulas and equations from the FEM:

i. Stiffness Matrix: An element's stiffness matrix relates the nodal displacements and nodal forces. A stiffness matrix for such a two-dimensional element may be demonstrated as follows:

$$[K_e] = \int_{\Omega_e} [B]^T [D] [B] d\Omega$$

where; $[K_e]$ is the component stiffness matrix, $[B]$ is the strain-displacement matrix, $[D]$ is the constitutive matrix, and Ω_e is the element domain (Barkanov, 2001).

ii. Global Stiffness Matrix: The component stiffness matrices of each element within the structure are combined to create the global stiffness matrix. A representation of the global stiffness matrix is as follows:

$$[K] = \sum_{e=1}^{N} [K_e]$$

where; $[K]$ N is the number of elements representing the global stiffness matrix.

iv. Nodal Force Vector: The nodal force vector represents the external loads placed on the structure. The following images illustrate the nodal force vector (Argyris et al., 1979):

$$\{f\} = \sum_{i=1}^{N} \{f_i\}$$

where; $\{f\}$ is the global

ACTIVITY 2.2.

How would you use a gyroscope to maintain the stability of a quadcopter drone during flight?

2.12. SUMMARY

The chapter on kinematics and rigid body dynamics covers several important concepts that are essential for understanding the motion of rigid bodies. The chapter starts with an overview of the aspects of kinematics, which is the study of motion without considering the forces that cause it. It then explains the concept of a frame of reference, which is a coordinate system used to describe the motion of an object. Next, the chapter introduces the concept of moment of inertia, which is a property of a body that determines its resistance to rotational motion. It also covers angular momentum and its conservation, which is a fundamental law in physics that states that the total angular momentum of an isolated system remains constant. The chapter also discusses the Euler equation, which describes the motion of a rigid body under the influence of external forces. It then covers the torque-free motion of symmetric rigid bodies, which is a type of motion that occurs when there are no external torques acting on the body.

The Lagrangian approach to rigid body dynamics is also discussed, which is a powerful method for solving problems related to the motion of rigid bodies. The chapter also covers gyroscopes, which are devices that are used to measure or maintain orientation. Finally, the chapter discusses numerical methods for solving problems related to rigid body motion. These methods are used to solve complex problems where analytical solutions are not feasible.

REVIEW QUESTIONS

1. What is kinematics and how is it related to other physical concepts?
2. Explain the concept of frame of reference and its importance in describing motion.
3. What is the moment of inertia and how does it relate to rotational motion?
4. How is angular momentum conserved and what are its applications in physics?
5. What is the Euler equation and how is it used in fluid dynamics?
6. Explain the concept of torque-free motion in rigid bodies.
7. What is Euler's equation and how is it used in rigid body dynamics?
8. How do we define angular momentum and what is its conservation law?
9. What is torque-free motion and how is it related to symmetric rigid body?
10. What is the frame of reference and why is it important in the study of rigid body motion?

MULTIPLE CHOICE QUESTIONS

1. What is kinematics?

a. The study of forces

b. The study of motion

c. The study of energy
d. The study of heat

2. **What is the moment of inertia?**
   a. The measure of a body's resistance to rotation
   b. The measure of a body's mass
   c. The measure of a body's velocity
   d. The measure of a body's potential energy
3. **What is angular momentum?**
   a. The measure of a body's resistance to linear motion
   b. The measure of a body's velocity
   c. The measure of a body's rotation
   d. The measure of a body's acceleration
4. **What is the Euler equation?**
   a. An equation used to describe fluid dynamics
   b. An equation used to describe rigid body motion
   c. An equation used to describe linear motion
   d. An equation used to describe circular motion
5. **What is torque-free motion?**
   a. Motion with constant torque
   b. Motion with varying torque
   c. Motion without any torque
   d. Motion without any velocity
6. **What is the Lagrangian approach to rigid body dynamics?**
   a. An analytical method used to describe the motion of rigid bodies
   b. A numerical method used to describe the motion of rigid bodies
   c. A graphical method used to describe the motion of rigid bodies
   d. A conceptual method used to describe the motion of rigid bodies
7. **What are gyroscopes?**
   a. Devices used to measure torque
   b. Devices used to measure velocity
   c. Devices used to measure angular momentum
   d. Devices used to maintain orientation
8. **What are numerical methods used for in kinematics?**
   a. To calculate the velocity of a moving object
   b. To calculate the acceleration of a moving object

c. To solve complex equations related to motion
d. To determine the force acting on a moving object

Answers to Multiple Choice Questions

1. (b); 2. (a); 3. (c); 4. (a); 5. (c); 6. (a); 7. (d); 8. (c)

REFERENCES

1. Adlaj, S., (2021). Torque Free Motion of a Rigid Body: From Feynman Wobbling Plate to Dzhanibekov Flipping Wingnut (No. 6011). Easy Chair.
2. Alsing, P. M., McMahon, D., & Milburn, G. J., (2004). Teleportation in a non-inertial frame. Journal of Optics B: Quantum and Semiclassical Optics, 6(8), S834.
3. Amendola, L., (2003). Acceleration at z> 1? Monthly Notices of the Royal Astronomical Society, 342(1), 221–226.
4. Anastassi, Z. A., & Simos, T. E., (2005). An optimized Runge–Kutta method for the solution of orbital problems. Journal of Computational and Applied Mathematics, 175(1), 1–9.
5. Argyris, J. H., Balmer, H., Doltsinis, J. S., Dunne, P. C., Haase, M., Kleiber, M., & Scharpf, D. W., (1979). Finite element method—The natural approach. Computer Methods in Applied Mechanics and Engineering, 17, 1–106.
6. Armstrong, H. L., (1967). On the precession and nutation of gyroscopes. American Journal of Physics, 35(9), 883–885.
7. Attanasio, O. P., & Low, H., (2004). Estimating Euler equations. Review of Economic Dynamics, 7(2), 406–435.
8. Babuška, I., & Suri, M., (1990). The p-and hp versions of the finite element method, an overview. Computer Methods in Applied Mechanics and Engineering, 80(1–3), 5–26.
9. Babuška, I., (1973). The finite element method with penalty. Mathematics of Computation, 27(122), 221–228.
10. Babuska, I., Szabo, B. A., & Katz, I. N., (1981). The p-version of the finite element method. SIAM Journal on Numerical Analysis, 18(3), 515–545.
11. Bächer, M., Whiting, E., Bickel, B., & Sorkine-Hornung, O., (2014). Spin-it: Optimizing moment of inertia for spinnable objects. ACM Transactions on Graphics (TOG), 33(4), 1–10.
12. Bacry, H., & Lévy Leblond, J. M., (1968). Possible kinematics. Journal of Mathematical Physics, 9(10), 1605–1614.
13. Barkanov, E., (2001). Introduction to the Finite Element Method (pp. 1–70). Institute of Materials and Structures Faculty of Civil Engineering Riga Technical University.

14. Becattini, F., Piccinini, F., & Rizzo, J., (2008). Angular momentum conservation in heavy ion collisions at very high energy. Physical Review C, 77(2), 024906.
15. Becker, M., Tessendorf, H., & Teschner, M., (2009). Direct forcing for Lagrangian rigid-fluid coupling. IEEE Transactions on Visualization and Computer Graphics, 15(3), 493–503.
16. Beliaev, S. T., (1961). Concerning the calculation of the nuclear moment of inertia. Nuclear Physics, 24(2), 322–325.
17. Benson, D. J., & Hallquist, J. O., (1986). A simple rigid body algorithm for structural dynamics programs. International Journal for Numerical Methods in Engineering, 22(3), 723–749.
18. Bergh, R., Lefevre, H., & Shaw, H., (1984). An overview of fiber-optic gyroscopes. Journal of Lightwave Technology, 2(2), 91–107.
19. Billat, L. V., & Koralsztein, J. P., (1996). Significance of the velocity at VO2max and time to exhaustion at this velocity. Sports Medicine, 22, 90–108.
20. Biral, F., Bertolazzi, E., & Bosetti, P., (2016). Notes on numerical methods for solving optimal control problems. IEEJ Journal of Industry Applications, 5(2), 154–166.
21. Bischoff, P. H., & Gross, S. P., (2011). Equivalent moment of inertia based on integration of curvature. Journal of Composites for Construction, 15(3), 263–273.
22. Bordoloi, A. D., & Variano, E., (2017). Rotational kinematics of large cylindrical particles in turbulence. Journal of Fluid Mechanics, 815, 199–222.
23. Böttcher, D., & Henk, J., (2012). Significance of nutation in magnetization dynamics of nanostructures. Physical Review B, 86(2), 020404.
24. Bovolo, F., & Bruzzone, L., (2006). A theoretical framework for unsupervised change detection based on change vector analysis in the polar domain. IEEE Transactions on Geoscience and Remote Sensing, 45(1), 218–236.
25. Braun, J., & Sambridge, M., (1995). A numerical method for solving partial differential equations on highly irregular evolving grids. Nature, 376(6542), 655–660.
26. Brill, D. R., & Cohen, J. M., (1966). Rotating masses and their effect on inertial frames. Physical Review, 143(4), 1011.
27. Brody, H., (1985). The moment of inertia of a tennis racket. The Physics Teacher, 23(4), 213–216.
28. Butcher, J. C., (1976). On the implementation of implicit Runge-Kutta methods. BIT Numerical Mathematics, 16(3), 237–240.
29. Butcher, J. C., (1996). A history of Runge-Kutta methods. Applied Numerical Mathematics, 20(3), 247–260.
30. Butikov, E., (2006). Precession and nutation of a gyroscope. European Journal of Physics, 27(5), 1071.
31. Carpenter, M. D., & Peck, M. A., (2009). Reducing base reactions with gyroscopic actuation of space-robotic systems. IEEE Transactions on Robotics, 25(6), 1262–1270.

32. Chorin, A. J., (1997). A numerical method for solving incompressible viscous flow problems. Journal of Computational Physics, 135(2), 118–125.

33. Ciufolini, I., (2007). Dragging of inertial frames. Nature, 449(7158), 41–47.

34. Civicioglu, P., (2012). Transforming geocentric cartesian coordinates to geodetic coordinates by using differential search algorithm. Computers & Geosciences, 46, 229–247.

35. Cooper, G., (1987). Stability of Runge-Kutta methods for trajectory problems. IMA Journal of Numerical Analysis, 7(1), 1–13.

36. Courtial, J., Dholakia, K., Allen, L., & Padgett, M. J., (1997). Second-harmonic generation and the conservation of orbital angular momentum with high-order Laguerre-Gaussian modes. Physical Review A, 56(5), 4193.

37. D'Aspremont, A., Scieur, D., & Taylor, A., (2021). Acceleration methods. Foundations and Trends® in Optimization, 5(1, 2), 1–245.

38. Davey, A., (1973). A simple numerical method for solving Orr–Sommerfeld problems. The Quarterly Journal of Mechanics and Applied Mathematics, 26(4), 401–411.

39. Delacroix, B., Rastoueix, J., Fradette, L., Bertrand, F., & Blais, B., (2021). CFD-DEM simulations of solid-liquid flow in stirred tanks using a non-inertial frame of reference. Chemical Engineering Science, 230, 116137.

40. Delgado-García, G., Vanrenterghem, J., Ruiz-Malagón, E. J., Molina-García, P., Courel-Ibáñez, J., & Soto-Hermoso, V. M., (2021). IMU gyroscopes are a valid alternative to 3D optical motion capture system for angular kinematics analysis in tennis. Proceedings of the Institution of Mechanical Engineers, Part P: Journal of Sports Engineering and Technology, 235(1), 3–12.

41. Di Gregorio, R., (2004). Kinematics of the translational 3-URC mechanism. J. Mech. Des., 126(6), 1113–1117.

42. Duda, K. R., & Newman, D. J., (2013). Variable vector countermeasure suit (V2Suit) for space exploration. In: 2013 IEEE Aerospace Conference (pp. 1–8). IEEE.

43. Fadeev, P., Timberlake, C., Wang, T., Vinante, A., Band, Y. B., Budker, D., & Kimball, D. F. J., (2021). Ferromagnetic gyroscopes for tests of fundamental physics. Quantum Science and Technology, 6(2), 024006.

44. Fang, J., Ren, Y., & Fan, Y., (2013). Nutation and precession stability criterion of magnetically suspended rigid rotors with gyroscopic effects based on positive and negative frequency characteristics. IEEE Transactions on Industrial Electronics, 61(4), 2003–2014.

45. Fish, J., (1992). The s-version of the finite element method. Computers & Structures, 43(3), 539–547.

46. Folgueira, M., & Souchay, J., (2001). A numerical model of the torque-free motion. To the ICRS and J2000, 67, 202.

47. Fuhrer, J. C., & Rudebusch, G. D., (2004). Estimating the Euler equation for output. Journal of Monetary Economics, 51(6), 1133–1153.

48. Galdi, G. P., & Silvestre, A. L., (2002). Strong solutions to the problem of motion of a rigid body in a Navier—Stokes liquid under the action of prescribed forces and torques. Nonlinear Problems in Mathematical Physics and Related Topics I: In Honor of Professor OA Ladyzhenskaya, 121–144.
49. Ge, Z. M., & Yu, J. K., (2000). Pragmatical asymptotical stability theorems on partial region and for partial variables with applications to gyroscopic systems. Journal of Mechanics, 16(4), 179–187.
50. Gerdan, G. P., & Deakin, R. E., (1999). Transforming cartesian coordinates X, Y, Z to geographical coordinates ϕ, λ, h. Australian Surveyor, 44(1), 55–63.
51. Giles, M. B., (1990). Nonreflecting boundary conditions for Euler equation calculations. AIAA Journal, 28(12), 2050–2058.
52. Giżynski, K., Lee, T., & Grzybowski, B. A., (2017). Dynamic self assembly of magnetic/polymer composites in rotating frames of reference. Advanced Materials, 29(33), 1700614.
53. Glowinski, R., Pan, T. W., & Periaux, J., (1998). Distributed Lagrange multiplier methods for incompressible viscous flow around moving rigid bodies. Computer Methods in Applied Mechanics and Engineering, 151(1, 2), 181–194.
54. Gonzalez-Longatt, F., Roldan-Fernandez, J. M., Chamorro, H. R., Arnaltes, S., & Rodriguez- Gonzalez-Longatt, F., Roldan-Fernandez, J. M., Chamorro, H. R., Arnaltes, S., & Rodriguez-Amenedo, J. L. (2021). Investigation of inertia response and rate of change of frequency in low rotational inertial scenario of synchronous dominated system. Electronics, 10(18), 2288.
55. Gu, D., El-Sheimy, N., Hassan, T., & Syed, Z., (2008). Coarse alignment for marine SINS using gravity in the inertial frame as a reference. In: 2008 IEEE/ION Position, Location and Navigation Symposium (pp. 961–965). IEEE.
56. Hairer, E., (1978). A Runge-Kutta method of order 10. IMA Journal of Applied Mathematics, 21(1), 47–59.
57. Han, B., Chen, Y., Li, M., Zheng, S., & Zhang, X., (2020). Stable control of nutation and precession for the radial four-degree-of-freedom AMB-rotor system considering strong gyroscopic effects. IEEE Transactions on Industrial Electronics, 68(11), 11369–11378.
58. Han, B., Post, D., & Ifju, P., (2001). Moiré interferometry for engineering mechanics: Current practices and future developments. The Journal of Strain Analysis for Engineering Design, 36(1), 101–117.
59. Hill, R., (1962). Acceleration waves in solids. Journal of the Mechanics and Physics of Solids, 10(1), 1–16.
60. Horwitz, J. L., Ho, C. W., Scarbro, H. D., Wilson, G. R., & Moore, T. E., (1994). Centrifugal acceleration of the polar wind. Journal of Geophysical Research: Space Physics, 99(A8), 15051–15064.
61. Istomin, Y. N., & Gunya, A. A., (2020). Centrifugal acceleration of protons by a supermassive black hole. Monthly Notices of the Royal Astronomical Society, 492(4), 4884–4891.

62. Jagota, V., Sethi, A. P. S., & Kumar, K., (2013). Finite element method: An overview. Walailak Journal of Science and Technology (WJST), 10(1), 1–8.

63. Johnson, S. C., Papageorgiou, D. J., Mallapragada, D. S., Deetjen, T. A., Rhodes, J. D., & Webber, M. E., (2019). Evaluating rotational inertia as a component of grid reliability with high penetrations of variable renewable energy. Energy, 180, 258–271.

64. Kadalbajoo, M. K., & Gupta, V., (2010). A brief survey on numerical methods for solving singularly perturbed problems. Applied Mathematics and Computation, 217(8), 3641–3716.

65. Kinoshita, H., (1992). Analytical expansions of torque-free motions for short and long axis modes. Celestial Mechanics and Dynamical Astronomy, 53, 365–375.

66. Königsberger, M., Hlobil, M., Delsaute, B., Staquet, S., Hellmich, C., & Pichler, B., (2018). Hydrate failure in ITZ governs concrete strength: A micro-to-macro validated engineering mechanics model. Cement and Concrete Research, 103, 77–94.

67. Kraaijevanger, J. F. B. M., (1991). Contractivity of Runge-Kutta methods. BIT Numerical Mathematics, 31(3), 482–528.

68. Küchemann, S., Klein, P., Fouckhardt, H., Gröber, S., & Kuhn, J., (2020). Students' understanding of non-inertial frames of reference. Physical Review Physics Education Research, 16(1), 010112.

69. Lamy, P. L., & Burns, J. A., (1972). Geometrical approach to torque free motion of a rigid body having internal energy dissipation. American Journal of Physics, 40(3), 441–445.

70. Laube, P., Imfeld, S., & Weibel, R., (2005). Discovering relative motion patterns in groups of moving point objects. International Journal of Geographical Information Science, 19(6), 639–668.

71. Lauer, T. R., & Postman, M., (1994). The motion of the local group with respect to the 15,000 kilometer per second Abell cluster inertial frame. Astrophysical Journal, Part 1 (ISSN 0004-637X), 425(2), 418–438.

72. Liu, P. L., & Der, K. A., (1986). Multivariate distribution models with prescribed marginals and covariances. Probabilistic Engineering Mechanics, 1(2), 105–112.

73. Loarie, S. R., Duffy, P. B., Hamilton, H., Asner, G. P., Field, C. B., & Ackerly, D. D., (2009). The velocity of climate change. Nature, 462(7276), 1052–1055.

74. Logg, A., (2007). Automating the finite element method. Archives of Computational Methods in Engineering, 14, 93–138.

75. Lum, K. Y., Bernstein, D. S., & Coppola, V. T., (1995). Global stabilization of the spinning top with mass imbalance. Dynamics and Stability of Systems, 10(4), 339–365.

76. Michelson, A. A., & Morley, E. W., (1887). On the relative motion of the earth and the luminiferous ether. American Journal of Science, 3(203), 333–345.

77. Michelson, A. A., (1904). LXXII. Relative motion of earth and æther. The London, Edinburgh, and Dublin Philosophical Magazine and Journal of Science, 8(48), 716–719.

78. Misslisch, H., Tweed, D., Fetter, M., Sievering, D., & Koenig, E., (1994). Rotational kinematics of the human vestibulo-ocular reflex. III. Listing's law. Journal of Neurophysiology, 72(5), 2490–2502.

79. Muhanna, R. L., Zhang, H., & Mullen, R. L., (2007). Interval finite elements as a basis for generalized models of uncertainty in engineering mechanics. Reliable Computing, 13, 173–194.

80. Nabae, A., & Tanaka, T., (1996). A new definition of instantaneous active-reactive current and power based on instantaneous space vectors on polar coordinates in three-phase circuits. IEEE Transactions on Power Delivery, 11(3), 1238–1243.

81. Naganathan, S. R., Fürthauer, S., Nishikawa, M., Jülicher, F., & Grill, S. W., (2014). Active torque generation by the actomyosin cell cortex drives left–right symmetry breaking. Elife, 3, e04165.

82. Niu, X., Zhang, Q., & Yanagidani, T., (2013). High-rate precise point positioning (PPP) to measure seismic wave motions: An experimental comparison of GPS PPP with inertial measurement units. Journal of Geodesy, 87, 361–372.

83. Novikov, S. P., & Shmel'tser, I., (1982). Periodic solutions of Kirchhoff's equations for the free motion of a rigid body in a fluid and the extended theory of Lyusternik-Shnirel'man &-morse (LSM). I. Functional Analysis and its Applications, 15(3), 197–207.

84. Ogbu, J. U., (1993). Differences in cultural frame of reference. International Journal of Behavioral Development, 16(3), 483–506.

85. Palmer, A. G., & Massi, F., (2006). Characterization of the dynamics of biomacromolecules using rotating-frame spin relaxation NMR spectroscopy. Chemical Reviews, 106(5), 1700–1719.

86. Perea, A., Castellano, J., Alday, L., & Hernández-Mendo, A., (2012). Analysis of behavior in sports through polar coordinate analysis with MATLAB®. Quality & Quantity, 46, 1249–1260.

87. Pichler, B., & Hellmich, C., (2011). Upscaling quasi-brittle strength of cement paste and mortar: A multi-scale engineering mechanics model. Cement and Concrete Research, 41(5), 467–476.

88. Pina, E., (2017). Jacobi's approach to the Lagrange top. European Journal of Physics, 38(3), 035001.

89. Robertson, H. P., (1935). Kinematics and world-structure. The Astrophysical Journal, 82, 284.

90. Robertson, H. P., (1936). Kinematics and world-structure II. The Astrophysical Journal, 83, 187.

91. Rogers, K. B., & Kimpston, R. D., (1992). Acceleration: What we do vs. what we know. Educational Leadership, 50(2), 58–61.

92. Romano, M., (2008). Exact analytic solutions for the rotation of an axially symmetric rigid body subjected to a constant torque. Celestial Mechanics and Dynamical Astronomy, 101, 375–390.

93. Saha, S., Mishra, M. K., Reddy, C. M., & Desiraju, G. R., (2018). From molecules to interactions to crystal engineering: Mechanical properties of organic solids. Accounts of Chemical Research, 51(11), 2957–2967.
94. Salisbury, J. K., (1980). Active stiffness control of a manipulator in cartesian coordinates. In: 1980 19th IEEE Conference on Decision and Control Including the Symposium on Adaptive Processes (pp. 95–100). IEEE.
95. Simmons, A. J., & Burridge, D. M., (1981). An energy and angular-momentum conserving vertical finite-difference scheme and hybrid vertical coordinates. Monthly Weather Review, 109(4), 758–766.
96. Söderkvist, J., (1994). Micromachined gyroscopes. Sensors and Actuators A: Physical, 43(1–3), 65–71.
97. Sorensen, R. A., (1973). Nuclear moment of inertia at high spin. Reviews of Modern Physics, 45(3), 353.
98. Southern, W. T., & Jones, E. D., (2004). Types of acceleration: Dimensions and issues. A Nation Deceived: How Schools Hold Back America's Brightest Students, 2, 5–12.
99. Takagi, S., (1991). Quantum dynamics and non-inertial frames of reference. I: Generality. Progress of Theoretical Physics, 85(3), 463–479.
100. Tan, C. W., Park, S., Mostov, K., & Varaiya, P., (2001). Design of gyroscope-free navigation systems. In: ITSC 2001. 2001 IEEE Intelligent Transportation Systems. Proceedings (Cat. No. 01TH8585) (pp. 286–291). IEEE.
101. Taner, M. T., & Koehler, F., (1969). Velocity spectra—Digital computer derivation applications of velocity functions. Geophysics, 34(6), 859–881.
102. Tanriverdi, V., (2021). Why Do We Use Constants of Motion While Studying the Motion of a Heavy Symmetric Top? arXiv preprint arXiv:2103.13984.
103. Thompson, H. B., (1967). Calculation of cartesian coordinates and their derivatives from internal molecular coordinates. The Journal of Chemical Physics, 47(9), 3407–3410.
104. Tong, K., & Granat, M. H., (1999). A practical gait analysis system using gyroscopes. Medical Engineering & Physics, 21(2), 87–94.
105. Tsai, L. W., Walsh, G. C., & Stamper, R. E., (1996). Kinematics of a novel three DOF translational platform. In: Proceedings of IEEE International Conference on Robotics and Automation (Vol. 4, pp. 3446–3451). IEEE.
106. Tweed, D. O. U. G. L. A. S., & Vilis, T., (1987). Implications of rotational kinematics for the oculomotor system in three dimensions. Journal of Neurophysiology, 58(4), 832–849.
107. Tyler, R. H., & Mysak, L. A., (1995). Electrodynamics in a rotating frame of reference with application to global ocean circulation. Canadian Journal of Physics, 73(5, 6), 393–402.
108. Vojtech, Š., Dušan, N., & Jozef, H., (2012). Calculation of robot position utilizing accelerometers in non-inertial frame of reference. In: 2012 ELEKTRO (pp. 373–376).

IEEE.

109. Wagner, J. F., (2018). About motion measurement in sports based on gyroscopes and accelerometers—An engineering point of view. Gyroscopy and Navigation, 9, 1–18.
110. Walborn, S. P., De Oliveira, A. N., Thebaldi, R. S., & Monken, C. H., (2004). Entanglement and conservation of orbital angular momentum in spontaneous parametric down-conversion. Physical Review A, 69(2), 023811.
111. Wilkie, D., Van, D. B. J., & Manocha, D., (2009). Generalized velocity obstacles. In: 2009 IEEE/RSJ International Conference on Intelligent Robots and Systems (pp. 5573–5578). IEEE.
112. Wittkowski, R., & Löwen, H., (2012). Self-propelled Brownian spinning top: Dynamics of a biaxial swimmer at low Reynolds numbers. Physical Review E, 85(2), 021406.
113. Wolfenstein, L., (1951). Conservation of angular momentum in the statistical theory of nuclear reactions. Physical Review, 82(5), 690.
114. Xu, P., Shu, Y., Niu, X., Liu, J., Yao, W., & Chen, Q., (2019). High-rate multi-GNSS attitude determination: Experiments, comparisons with inertial measurement units and applications of GNSS rotational seismology to the 2011 Tohoku Mw9. 0 earthquake. Measurement Science and Technology, 30(2), 024003.
115. Yetkin, H., Kalouche, S., Vernier, M., Colvin, G., Redmill, K., & Ozguner, U., (2014). Gyroscopic stabilization of an unmanned bicycle. In: 2014 American Control Conference (pp. 4549–4554). IEEE.
116. Zadravec, M., Basic, S., & Hribersek, M., (2007). The influence of rotating domain size in a rotating frame of reference approach for simulation of rotating impeller in a mixing vessel. Journal of Engineering Science and Technology, 2(2), 126–138.
117. Zennaro, M., (1986). Natural continuous extensions of Runge-Kutta methods. Mathematics of Computation, 46(173), 119–133.
118. Zhang, X., Zhu, W., Xie, Y., Wu, X., & Guo, J., (2023). Cross-coupling relative dynamics for on-orbit assembly in rotating frame of reference. Applied Mathematical Modeling, 116, 372–392.
119. Zhu, J. Z., & Zienkiewicz, O. C., (1988). Adaptive techniques in the finite element method. Communications in Applied Numerical Methods, 4(2), 197–204.

CHAPTER 3

STATIC AND EQUILIBRIUM

UNIT INTRODUCTION

The topic of Statics and Equilibrium is a major component in the field of mechanics that focuses on the movement and forces of things in the real world. The ideas of statics and equilibrium are critical for understanding structures, machinery, and other systems functioning at rest or in uniform motion. Statics is concerned with the study of objects which are in static or uniform motion. In contrast, equilibrium is the condition of a system that exists when the net force and net torque operating upon them are both zero (Ramos et al., 2016). This section will cover the fundamental ideas and principles for statics and stabilization, such as rigid body equilibrium, resistance, and many applications of all these ideas in diverse disciplines. Readers should have a strong knowledge of the concepts of statics and equilibrium through the end of this section and be capable of using these ideas to address real-world issues in physics, engineering, and other related subjects (Orlov, 2015) (Figure 3.1).

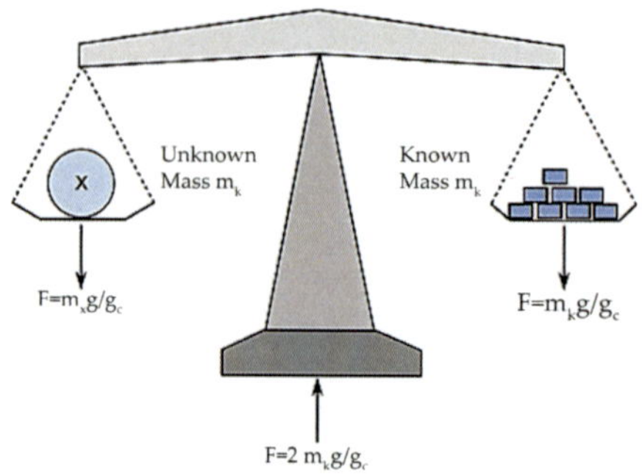

Figure 3.1. Illustration of static equilibrium.

Source: Science Direct, Creative Commons License.

Learning Objectives

At the end of this chapter, readers will be able to:

- Explain the importance of statics in engineering mechanics;
- Define equilibrium and its general conditions for rigid bodies;
- Describe the concept of center of gravity and its significance in statics;
- Discuss the laws of friction and friction angles.

Key Terms

- Center of gravity
- Coefficient of friction
- Engineering mechanics
- Equilibrium
- Force
- Friction
- Friction angles
- Kinetic friction
- Laws of friction
- Moment
- Rigid bodies
- Rolling friction
- Static equilibrium
- Statics
- Torque

3.1. STATICS

The analysis of bodies during rest or in a steady state is the focus of the mechanic's subfield, known as statics. It entails analyzing the periods and forces that act on an item and the circumstances under which a body falls into an equilibrium position. The focus of statics is the behavior of things that are not moving but rather at resting or moving with such a constant speed. Studying numerous physical variables, including forces, periods, and torque multiplication, is a key component in statics. A moment would be an effect created by force applied on an item, whereas a force is a pushing or pulling which interacts with an object. On the reverse hand, torque is a measurement of the rotating force acting on an item (Steif & Dollar, 2005).

The stability of an item is crucial to understanding statics. Whenever the net force and the net moment applied on an item are zero, that situation is called equilibrium. It indicates that the thing is either stationary or moving continuously (Tewary, 1973).

Many real-world applications of statics research may be found in engineering, physics, and other related disciplines. Statics is utilized, for instance, in designing and planning machines, including mechanical systems, and developing and evaluating constructions such as buildings, bridges, and other architectural structures. The research into the human organism and its motions and geological occurrences, including landslides and earthquakes, require statics (Lerner, 1937).

3.2. IMPORTANCE OF STATICS

The focus of the crucial mechanic's subdiscipline known as statics is the study of things at rest or moving uniformly—some main justifications for statics' significance (Boroudjerdi et al., 2005).

3.2.1. Understanding Structures

Knowing the structure of objects like bridges, buildings, and towers requires knowledge of statics. It aids engineers in predicting how all these structures will perform under various circumstances and loads and how to create secure and stable designs (Lopez et al., 1998).

3.2.2. Designing Machines

The fundamentals of statics are essential to efficiently operating many machinery and physical systems. Statics must be understood to analyze and develop devices like engines, pumps, and turbines (Stiny & March, 1981).

3.2.3. Preventing Accidents

By recognizing possible risks and dangers in a specific method, statics may assist in averting accidents. Engineering professionals may create safer systems that are less prone to malfunction or failure by studying the momentary and permanent forces operating on a system (Wachter & Yorio, 2014).

3.2.4. Enhancing Performance

Studying statics may help a system operate better while also assuring safety. Engineers may create more effective and efficient structures by optimizing the moments and forces operating on a system (Medlin & Green, 2009).

3.2.5. Advancing Research

Several study areas, including geophysics, biomechanics, and materials science, depend heavily on statics. Researchers may create new materials and technologies by studying statics and obtaining new insights into how these systems behave (Battilana & Lee, 2014). Statics plays a crucial role in mechanics analysis and has several applications within engineering, science, and other disciplines. It is crucial for furthering innovation and study and guaranteeing the functionality and safety of buildings and systems (Raudenbush, 2008) (Figure 3.2).

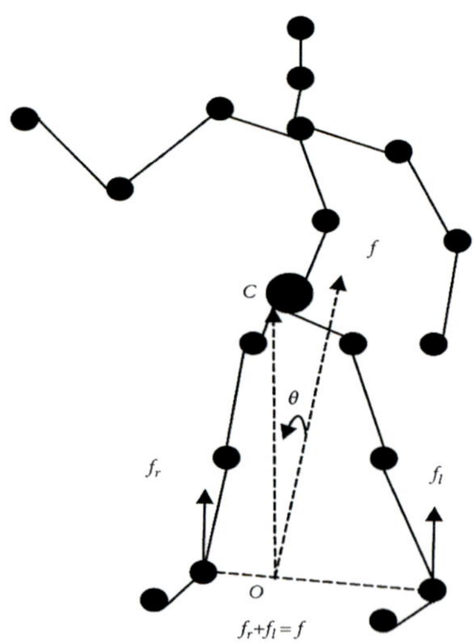

Figure 3.2. Diagram of human body mechanics analysis.

Source: Yating Wei, Creative Commons License.

3.3. EQUILIBRIUM

A system is in equilibrium when it has reached a stable, unchanging, and balanced condition. Equilibrium is frequently used to describe the condition of a system where the net torque and net force operating mostly on the system are zero in engineering and physics. In all other words, neither the system is spinning nor accelerating (Eyster & Rabin, 2005).

3.3.1. Types of Equilibrium

Dynamic and static equilibrium are the two primary categories of equilibrium (Zagare, 1984).

3.3.1.1. Static Equilibrium

It happens when an item is stationary or moving at a steady speed. As a result, the item is not accelerated, and no net force acts on it. If all forces that act on an item are still in equilibrium or cancel out the other, a condition of balance known as an equilibrium state could be reached (Kruskal & Oberman, 1958).

3.3.1.2. Dynamic Equilibrium

Conversely, dynamic equilibrium happens when an item travels in a circle at a constant speed. Even though no net force is exerted on the object in this situation, the object accelerates because of its direction shift. It indicates that overall velocity changes, although the object's speed remains constant (Smith & Lewis, 2011).

Did you Know?

Equilibrium is a state where all forces acting on an object balance out, resulting in no change in motion. Understanding equilibrium is important in the design of structures and mechanical systems, and it is a fundamental concept in physics.

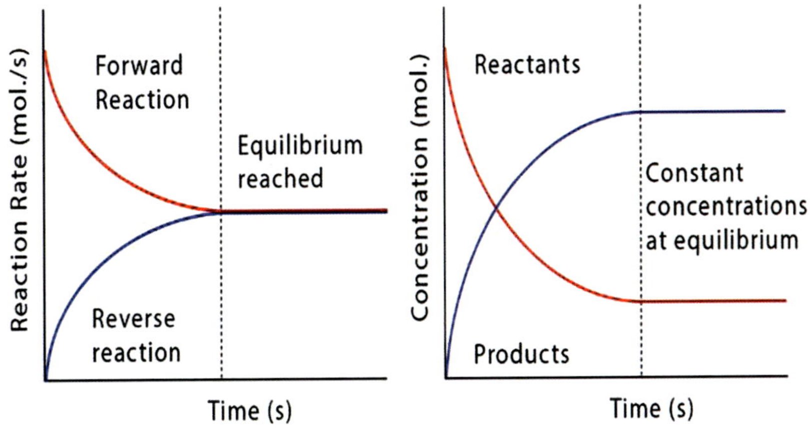

Source: Chemistry Learner, Creative Commons License.

Figure 3.3. Example of dynamic equilibrium.

In engineering, physics, and other disciplines, equilibrium is crucial because it clarifies how systems and things behave in various situations. The design of buildings, machinery, and some other systems, along with the study of natural processes like the movement of stars and planets, may all benefit from understanding such conditions necessary for equilibrium (Bernadó & Blackledge, 2010) (Figure 3.3).

3.3.2. Rigid Body in Motion

One must comprehend a rigid body's motions to comprehend its balance. A rigid body can move in two ways (Morgul, 1991).

3.3.2.1. Translational Motion

The rigid bodies based on another movement in which each constituent component moves linearly. In contrast, a perfect translational motion involves a rigid body that flows in a straight path and has all its particles always moving at the same speed. Consider pulling a box across a flat surface. Each of the box's components is traveling in the same way with the same speed at any given time because it is going down a straight line with no sideways movement. Under this scenario, the rigid body, in this instance, the box, is only in translational motion (Lawton, 1983).

Due to the spherical form of the ball, all of the particles would not move across the linear plane with the same speed at any one time, yet they will nonetheless roll across a single surface. That ball is not moving in a straight line during this action. What transpires when a stiff body can only move linearly or in translation? The second type of rotational motion is the solution (Kinoshita, 2009).

3.3.2.2. Rotational Motion

Whenever a rigid body remains fixed along one line or point, only one motion it is capable of is rotation from around a fixed point, which is what rotational motion is. The rotational axis is a rigid body's fixed line upon which it rotates. The rotations of an earth's axis or even the movement of a fan are the most typical examples of this motion. In rotations, a rigid body moves in a circle centered on the axis while all of its constituent components move (Schuler et al., 1967).

The ball goes forward in a linear plane and spins with only a fixed point when rolling mostly on the ground. The ball components move ahead concretely, and a rotational movement is centered at the axis. Therefore, the ball moves both in translations and rotations in this situation. A rigid body either moves in translational motion or combines translational and rotational movement while in movement but not fixed to a location. If an object moves to a point or line while being fixed, it is rotating (Coplan, 1989).

3.4. GENERAL CONDITIONS FOR EQUILIBRIUM OF RIGID BODY

Scientists say it has reached equilibrium whenever a rigid body's angular and linear acceleration are zero concerning an inertial reference point, sci. An object in equilibrium could thus move, but its angular and linear velocities should remain constant. Whenever a rigid body is in motion in the chosen frame of reference, researchers say that it has reached static equilibrium. It should be noted that an object could be at rest in our chosen perspective but appear to be operating in constant acceleration with constant speed to a viewer moving compared to our frame of reference once at a constant speed (Schlanbusch et al., 2012). The distinction between a nation of uniform circular motion and a nation of the remainder is artificial. Conversely, what appears to be in equilibrium to us becomes in dynamic equilibrium to a moving viewer because movement is relative. There is no difference between static balance and equilibrium in an inertial frame of reference since the physics principles remain the same for all inertial reference frames (Takashi & Hughes, 1992).

Newton's second rule of motion states that a rigid body's linear acceleration is generated by either a force applied that affects the body, or

$$\sum_k \vec{\mathbf{F}}_k = m\vec{\mathbf{a}}_{\mathrm{CM}}. \tag{3.1}$$

Here, the body is being influenced by all external influences together, where m is its mass and $\vec{\mathbf{a}}_{\mathrm{CM}}$ is the linear acceleration of its center of mass. The linear acceleration becomes zero (0) when everything is balanced. Eqn. (3.1) yields the following equation if we set the constant acceleration to zero (Wang & Xu, 2013).

3.4.1. First Equilibrium Condition

The first equilibrium position for the equilibrium point of a rigid body expresses translational equilibrium:

$$\sum_k \vec{\mathbf{F}}_k = \vec{\mathbf{0}}. \tag{3.2}$$

Eqn. (3.2), a first equilibrium state, is the system must be able for forces, which we came across while examining how Newton's laws were applied.

This vector equation is equivalent to the following three scalar equations for the components of the net force (Parker & Valocchi, 1986):

$$\sum_k F_{kx} = 0, \quad \sum_k F_{ky} = 0, \quad \sum_k F_{kz} = 0. \tag{3.3}$$

Using Eqn. (3.3) as a comparison, we can say that the rotational acceleration $\vec{\alpha}$ Using Eqn. (3.3) as a comparison, we can say that such rotational acceleration.

$$\sum_k \vec{\tau}_k = I\vec{\alpha} \tag{3.4}$$

Here I am Using Eqn. (3.3) as a comparison; we can say that such rotational acceleration $\vec{\tau}_k$ of external forces in Eqn. (3.2). In equilibrium, the rotational acceleration is zero. The second equilibrium state is obtained by zeroing out the right side of Eqn. (3.4) (Litao et al., 2009):

3.4.2. Second Equilibrium Condition

The second equilibrium condition for the static equilibrium of a rigid body expresses rotating equilibrium:

$$\sum_k \vec{\tau}_k = \vec{0}. \tag{3.5}$$

The equilibrium point for torques we encountered when studying rotational dynamics is Eqn. (3.5), the second system that must be able. It is important to note that such an equilibrium equation may describe rotational equilibrium around any rotational axis (fixed or otherwise). Again, for such a position vector of a net torque, each vector equation is equal to three scalar equations (Watling, 2002):

$$\sum_k \tau_{kx} = 0, \quad \sum_k \tau_{ky} = 0, \quad \sum_k \tau_{kz} = 0. \tag{3.6}$$

According to the second equilibrium position, nothing net external torque may rotate an object around any axis when it is in equilibrium.

The initial and final equilibrium conditions are provided in a certain reference frame. The first criterion is independent of a source of the frame of reference because it only includes forces. The second requirement, however, is related to torque, which is a crossing product., $\vec{\tau}_k = \vec{\mathbf{r}}_k \times \vec{\mathbf{F}}_k$, where the position vector

$\vec{\mathbf{r}}_k$ is the place where the applied force is included about an axis of rotation. Hence, the position of the axis inside the reference frame affects torque. The net torque about every rotated axis remains zero, though, since while the translational and rotational equilibrium conditions apply simultaneously in one reference point, they also hold regardless of the inertial reference frame. It's easy to understand why this is the case (Melander et al., 1984).

Suppose vector $R^{\rightarrow}$ is where a new inertial frame of reference's origin is located. S' in the old inertial frame of reference S. From our analysis of relative movement, we understand that in the new frame of reference S,' the position vector $\vec{\mathbf{r}}'_k$ of the point somewhere the $\vec{\mathbf{F}}_k$ is utilized is related to/via the equation (Dickson et al., 2016).

$$\vec{\mathbf{r}}'_k = \vec{\mathbf{r}}_k - \vec{\mathbf{R}}$$

We can now add all torques. $\vec{\tau}'_k = \vec{\mathbf{r}}'_k \times \vec{\mathbf{F}}_k$ of all external factors in a new frame of reference, S':

$$\sum_k \vec{\tau}'_k = \sum_k \vec{\mathbf{r}}'_k \times \vec{\mathbf{F}}_k = \sum_k (\vec{\mathbf{r}}_k - \vec{\mathbf{R}}) \times \vec{\mathbf{F}}_k = \sum_k \vec{\mathbf{r}}_k \times \vec{\mathbf{F}}_k - \sum_k \vec{\mathbf{R}} \times \vec{\mathbf{F}}_k = \sum_k \vec{\tau}_k - \vec{\mathbf{R}} \times \sum_k \vec{\mathbf{F}}_k = \vec{\mathbf{0}}.$$

The fact that the initial term vanishes in equilibrium there in old reference point S, according to Eqn. (3.5), and the second term vanishes due to Eqn. (3.2), was used to conclude this line of reasoning. Given that both requirements for equilibrium apply in an inertial reference point S, we can show that its net torque for any such reference point S' is zero (Fang, & Parreiras, 2002). It has the practical effect of leaving us free to select any location as the source of the frame of reference whenever applying equilibrium criteria for a rigid body. The physical details of the issue we are trying to solve determine the reference frame we use. The mathematical expression of the equilibrium state could be complex in one reference point, yet it could be much easier to solve the same circumstances in another frame. A pivot point designates the starting point of a chosen reference point (Eremeyev & Altenbach, 2014).

Remember

The general conditions for equilibrium of a rigid body are that the net force acting on it must be zero and the net torque about any point must also be zero. These conditions are important in the analysis and design of structures and mechanical systems to ensure stability and safety.

The six scalar equations represent equilibrium circumstances in the broadest sense (Eqns. (3.3) and (3.6)). As discussed in this chapter, we may reduce the number of formulas to three for planar equilibrium situations involving rotation along a fixed axis (Kim et al., 2008). The accepted practice is to use a reference point in which the rotation's axis is the z-axis. With such an axis selection, all forces with non-zero torques are on the xy plane, and as a result, contributions to a net torque only come from the y and x components of external factors. As a result, the net torque only has a z-component. Hence, we get the three following equilibrium requirements for torques and forces for planar issues with both the axis rotating parallel to the xy plane (Fusco & Morini, 2012):

$$F_{1x} + F_{2x} + \cdots + F_{Nx} = 0 \quad (3.7)$$

$$F_{1y} + F_{2y} + \cdots + F_{Ny} = 0 \quad (3.8)$$

$$\tau_1 + \tau_2 + \cdots + \tau_N = 0 \quad (3.9)$$

The total comprises the torques produced by all N exterior forces that impact the body. While we removed the subscript z from Eqn. (3.9) to simplify the notation, it is important to note that the summing here is overall contributions along the z-axis, meaning the axis of rotation. In Eqn. (3.9), the z-component of torque τ⃗k from the force F⃗k is (Partridge et al., 2015).

$$\tau_k = r_k F_k \sin\theta \quad (3.10)$$

Fk is the force's magnitude, and rk is the force's lever arm length. The angle θ is the angle between vectors r⃗k and F⃗k measuring from vector r⃗k to vector F⃗k in the counterclockwise direction. Depending on the axis of rotation induced through this torque independently, we frequently quantify the torque's magnitude and assign either one positive (+)(+) or a negative (–), (–) meaning when utilizing Eqn. (3.10) (Zhang & Ge, 2004). Net torque in Eqn. (3.9) is the product of terms, every one of which is determined from Eqn.

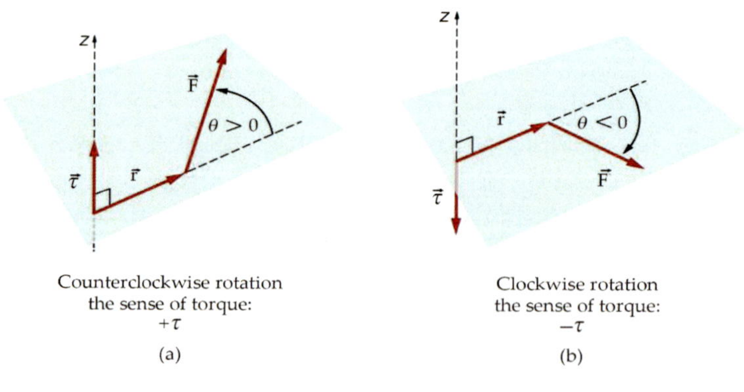

Figure 3.4. Torque of a force: (a) when a force's torque causes the axis of rotation to rotate counterclockwise, we say that the force's sense is positive, which denotes that the torque vector is parallel to the axis of rotation; and (b) whenever a force's torque causes the axis to rotate in a clockwise direction, we state that such force's sense is negative, which denotes that its torque vector is perpendicular to the axis of rotations.

Source: Lumen Learning, Creative Commons License.

(3.10) and should make sense. The ++ sign is applied to force elements in the +x-direction in Eqn. (3.7), whereas the —sign is applied to elements in the x-direction. While calculating force components along the y-axis in Eqn. (3.8), the same principle should be consistently used (Drummond et al., 1980) (Figure 3.4).

3.5. CENTER OF GRAVITY

In physics, the gravity center is a hypothetical location within a solid where, for the sake of convenience in some calculations, the whole weight of the body could be assumed to be concentrated. The idea might be helpful when planning static structures (like buildings and bridges) or forecasting how a moving body will behave when subjected to the effects of gravity (Caron et al., 1997).

The center of mass, as used by physicists, is the same as the center of gravity in a homogeneous gravitational field. Yet the two aren't necessarily related. For instance, the Moon's center of mass is very close to its geometric center (although it is not accurate since the Moon is not a completely uniform sphere), yet its center of gravity is somewhat shifted toward Earth due to the Moon's near side's higher gravitational pull (Horstmann & Dietz, 1990).

The geometric center of a body may also be where the body's center of gravity is located, particularly in symmetrically structured objects made of homogenous material. Therefore, it is likely that the center of gravity of an asymmetrical item made of several materials with different masses will be situated some distance from the object's geometric center. The center of gravity (or center of mass) may occasionally exist in space at a location independent of the physical substance, such as at the center of hollowed bodies or among the legs of an item with an irregular shape, including a tennis ball or even a chair (Grether & Mathys, 2010) (Figure 3.5).

TIPS

The center of gravity is the point where an object's weight is evenly distributed. To find it, suspend the object and mark the plumb lines, which intersect at the center of gravity. Understanding the center of gravity is important for stability analysis and safe system design.

The centers of gravity are published in charts and handbooks for the most popular geometric forms. The calculation will include adding the moments of weights of each component that makes the metallic plate around point A for a triangular metal frame like the one shown in the picture. The location of G concerning AC may be calculated by equating this total to the weight of a plate W multiplied by the unknown location of the center of gravitation G to AC. Integral calculus may calculate the moments' sum quickly and accurately (Manow et al., 2008).

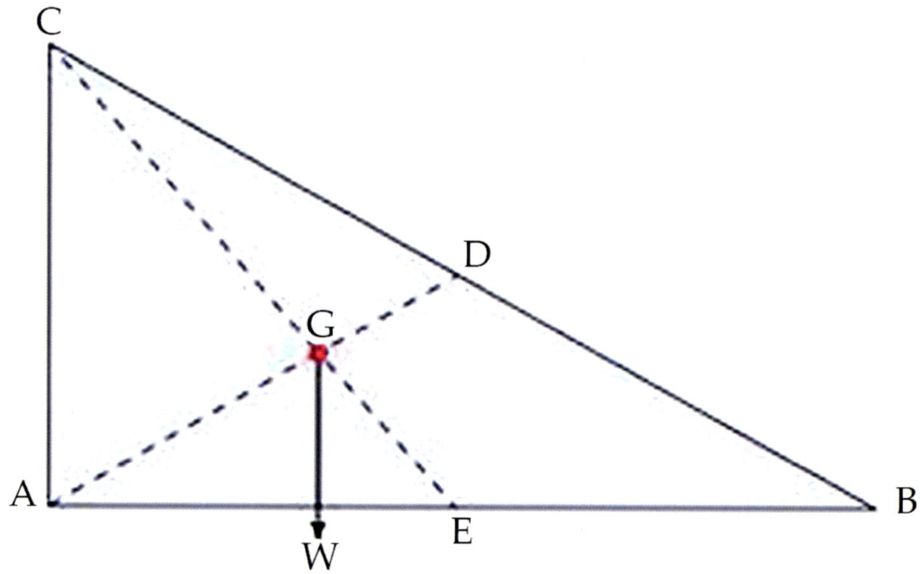

Figure 3.5. Diagram of center of gravity.

Source: Lumen Learning, Creative Commons License.

A straightforward physical process may also identify anyone's center of gravity. For instance, again, point G may be identified for the plate in the picture by hanging the plate by a string attached to point A before using a chord attached at C. A line AD is vertical whenever the plate is hung from A, while the line CE is vertical if it is suspended from C. There is a gravitational center at the point where AD and CE meet. The gravity center of any hung item is located just underneath that point (Lakhani & Mansfield, 2015).

3.6. FRICTION

The contact resistance one body exerts on another when the second body goes past the first body is known as friction. Friction is a force that always opposes a motion or an inclination to move (Blau, 2001).

When there is a propensity for one contacting area to slide over another surface, frictional force always increases in the opposite direction.

- We aim to reduce the friction forces' retarding effects on certain machines. Examples include multiple kinds of bearings, fluid flow, power screws, pipeline gears, aviation propulsion, and atmospheric missiles (Shinjo & Hirano, 1993).

- In some circumstances, we want to enhance the impact of friction on belt drives, brakes, wedges, and clutches (Suh et al., 1994).
- Friction between the foot and the ground is necessary for both the beginning and stopping of wheeled vehicles and normal walking (Urbakh et al., 2004).

3.6.1. Types of Friction

3.6.1.1. Dry Friction

Drying friction occurs whenever the unlubricated surfaces of two materials come into contact while moving or shows a tendency to slide. Coulomb friction is another name for this kind of friction (Shaw, 1986).

3.6.1.2. Fluid Friction

As neighboring layers of a fluid (liquid or gas) move at different speeds, fluid friction is created (Colebrook & White, 1937).

3.6.2. Mechanism of Friction

The contacting surface's smoothness is the main cause of friction. Think about a block with weight w perched on a flat surface. The contacting surface is indeed quite rough. Let P represent the horizontal force being applied; it will continually range from zero to a number big enough to move a block and keep it moving merely. The block's free body diagram displays active forces (such as the force applied P and the block's weight w) and reacting forces (i.e., normal reaction N and tangential frictional force F) (Bowden, 1952).

The notable characteristic of the frictional force F is that it will adapt until the limiting equilibrium position with a magnitude equivalent to the applied force P (Cook et al., 2013)

The graph in Figure 3.6 that shows applied force P vs. frictional force F might be used to visualize the concepts above (Braun & Naumovets, 2006).

Three different sorts of difficulties may now be identified by studying the chart. We have kinetic, static, and limiting friction in this situation (McNelley et al., 2008).

ACTIVITY 3.1.

How would you calculate the amount of force required to move a heavy object across a rough surface, taking into account the coefficient of friction between the object and the surface?

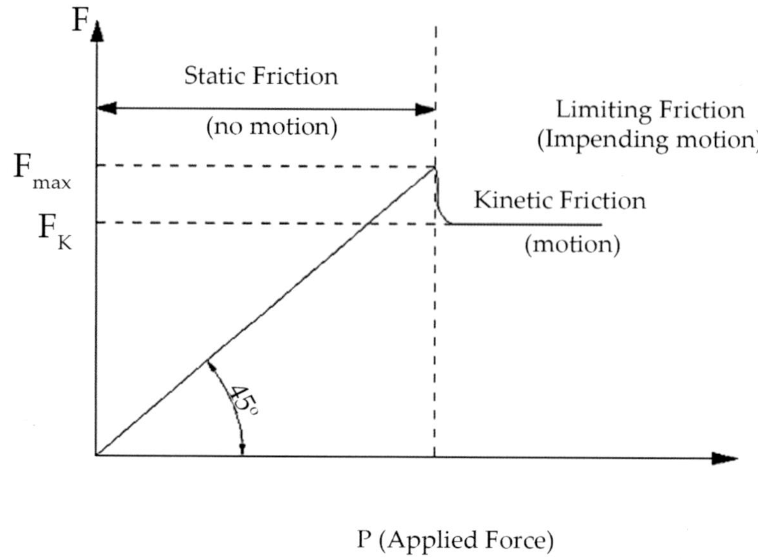

Figure 3.6. Graphical representation of applied force P vs. frictional force F.

Source: mvsrec, Creative Commons License.

3.6.2.1. Static Friction

When neither the state of approaching motion nor of motion exists in the problem, we must first assume equilibrium conditions and use F as the frictional force necessary to uphold the equilibrium point to calculate the actual force.

Here, we have three possibilities.

1. $F < F_{max}$ = The body is just in static equilibrium, which implies it is completely at rest.
2. $F = F_{max}$ = The body is in a limited equilibrium state, indicating imminent motion. And hence $F = F_{max} = \mu_s N$ is a valid equation.
3. $F > F_{max}$= When the body is in motion, $F = F_k = \mu_k N$ is a correct equation. The situation is unattainable since the surfaces cannot withstand any further force.

Then, for the maximal frictional force, as a result, the equilibrium assumption is flawed, and motion occurs (Johannes et al., 1973).

3.6.2.2. Limiting Friction

It is recognized that the situation of oncoming motion exists. In this case, an equilibrium body appears on the edge of sliding, indicating that the body has a limited equilibrium situation. "It denotes the highest magnitude of friction force that surface may generate on the block." as $F_{max.}$." It is mostly determined by the roughness of a surface's materials and the usual contact force these surfaces produce on one another (Carlson, 2006).

$F_{max} = \mu_s N$ is a valid equation.

3.6.2.3. Kinetic Friction

The contacting surfaces are understood to be in a state of relative motion. As a result, the body is already in motion. Kinetic friction gets placed $F_k = \mu_k N$ is a valid equation (Martins et al., 1990).

3.7. LAWS OF FRICTION

1. The frictional force has always been tangential to a contact area and operates reversely as the body moves.
2. The number of frictional forces self-adjusts to the applied force until the limited frictional force is achieved, at which point the body experiences approaching motion.
3. Limiting frictional force F_{max} is directly proportional to the typical reaction (Archard, 1957).

(Fmax = μsN).

4. For a moving body, the kinetic frictional force F_k is smaller than that of the limiting frictional force F_{max} and the relation $F_k = \mu_k N$ is applicable.
5. The amount of friction is determined by the substance in contact and the surface's smoothness (Popova & Popov, 2015).
6. The contact area on different surfaces has no bearing on the frictional force.
7. Body speed does not affect frictional force (Mo et al., 2009).
8. The friction coefficient at rest $\mu_{s.}$ is always bigger than the kinetic friction coefficient. μ_k (μ_k may be 25% smaller than μ_s in general) (Lincoln, 1953).

Note: $\mu_{s.}$ & μ_k are dimensionless.

3.8. FRICTION ANGLES

3.8.1. Angle of Friction

It refers to the angle created by limiting the frictional force's outcome. F_{max} together with the typical response N and the typical reactions (Ayininuola et al., 2009) (Figure 3.7).

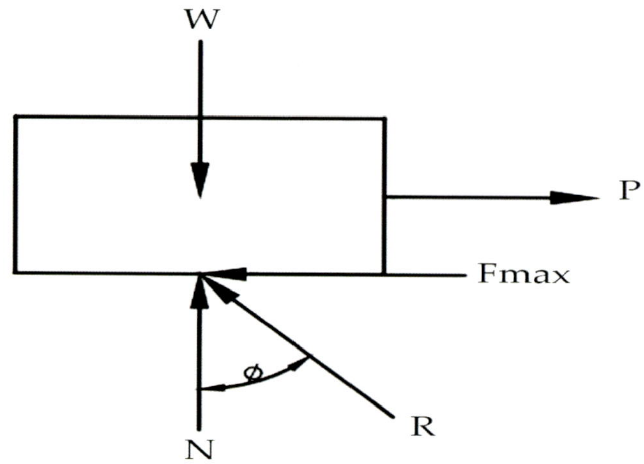

Figure 3.7. Illustration of F_{max} together with the typical response N and the typical reactions.

Source: mvsrec, Creative Commons License.

Think about the block having force p and weight W. The common reaction N would be parallel to the surface whenever the block is on the edge of motion. The limited frictional force F_{max} will operate in the reverse way as the force applied, as illustrated in Figure 3.7. The resulting reaction R, which operates at an angle to the typical reaction, can replace F_{max} and N. The friction angle is the name given to it (Schanz & Vermeer, 1996).

From Figure 3.7, we have:

F_{max} = R sinø μ_sN = R sinø (I)

(F_{max} = μ_sN) N= R cosø (II)

Dividing Eqn. (I) by Eqn. (II), we get:

Tanø = μ_s or ø = Tan^{-1} (μ_s)

3.8.2. Angle of Repose

It is the smallest angle at that a plane is inclined concerning the horizontal such that a body held at that angle slides downward on the plane without such application of every external force (due to self-weight) (Al-Hashemi et al., 2018).

As illustrated in the image, the block having weight W is perched on an inclined surface that forms an angle only with horizontal. The block would rest on the aircraft when it is tiny. A slope is attained when a block is poised to begin sliding when it is gradually increased. The angle of repose is the name for this angle (Van Burkalow, 1945).

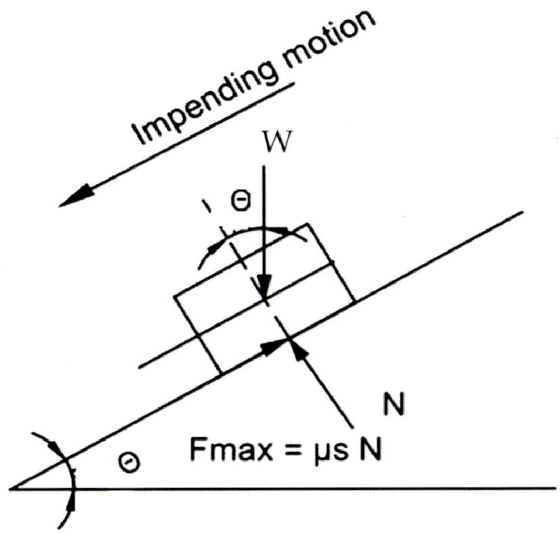

$\sum F_x = 0 \ \mu_s N - W \sin\Theta = 0$

$W\sin\Theta = \mu N \quad (I)$

$\sum F_y = 0$

$N - W \cos \Theta = 0$

$W\cos\Theta = N \quad (II)$

Dividing Eqn. (I) by Eqn. (II), we get:

$\tan \Theta = \mu_s$

In the previous discussion, we had $\tan ø = \mu_s$ which shows

The angle of friction ø = Angle of Repose Θ

The relationship mentioned above also demonstrates that such an angle of repose is based on an independent of body weight (Miura et al., 1997).

ACTIVITY 3.2

How would you analyze the forces acting on a ladder leaning against a wall to ensure that it is in static equilibrium and won't slip or tip over?

3.8.3. Cone of Friction

The angle of friction, which refers to the angle created by the resultant of limited friction forces with a normal response as indicated in Figure 3.7, is attained whenever the force applied P is adequate to generate the imminent motion of a given body (Jeon et al., 2015). If such applied force P's direction is progressively altered over a range of 3,600, the resulting R produces right circular cones with a semi-vertex angle. It is called the Cone of Friction (Fischer & Wriggers, 2006).

3.9. SUMMARY

The chapter on statics and equilibrium covers the study of forces and their effect on rigid bodies that are at rest or in a state of uniform motion. Statics is an important part of engineering mechanics as it helps in analyzing and designing structures that are safe and stable. Equilibrium refers to a state of balance where the net force and net moment on a body are both zero. The general conditions for equilibrium of rigid bodies include the sum of forces and sum of moments being equal to zero. The concept of center of gravity is significant in statics as it determines the stability of an object. It is the point where the entire weight of an object is assumed to act. Friction is the force that opposes motion between two surfaces in contact. The laws of friction state that there is a maximum force of friction that can act between two surfaces and that the force of friction is directly proportional to the normal force acting on the surfaces. The friction angle is the angle between the normal force and the force of friction. Kinetic friction refers to the force that opposes the motion of an object in motion, while static friction opposes the motion of an object at rest. The coefficient of friction is a value that represents the amount of friction between two surfaces and is calculated as the ratio of the force of friction to the normal force acting on the surfaces.

REVIEW QUESTIONS

1. What are statics and why is it important in engineering mechanics?
2. What is equilibrium and what are the general conditions for equilibrium of rigid bodies?
3. Describe the concept of center of gravity and its significance in statics.
4. What is friction and what are the laws of friction?
5. What is the friction angle and how does it relate to friction?
6. What is kinetic friction and how is it different from static friction?
7. What is the coefficient of friction and how is it calculated?
8. Explain the difference between rolling and sliding friction.

MULTIPLE CHOICE QUESTIONS

1. What are statics?

a. The study of motion
b. The study of forces
c. The study of equilibrium
d. The study of energy

2. Which of the following is an important factor in statics?

a. Acceleration

b. Velocity
c. Force
d. Displacement

3. **What are the general conditions for equilibrium of rigid bodies?**
a. Sum of forces and sum of moments equal zero
b. Sum of forces equal zero
c. Sum of moments equal zero
d. Sum of forces and moments equal to mass

4. **What is the significance of center of gravity in statics?**
a. Determines the motion of an object
b. Determines the direction of the force
c. Determines the stability of an object
d. Determines the speed of an object

5. **What are the laws of friction?**
a. Kinetic and potential friction
b. Rolling and sliding friction
c. Static and kinetic friction
d. Friction and non-friction

6. **What is the friction angle?**
a. The angle between the normal force and the force of friction
b. The angle between two forces
c. The angle of displacement
d. The angle of velocity

7. **What is kinetic friction?**
a. The force that opposes the motion of an object at rest
b. The force that opposes the motion of an object in motion
c. The force that causes an object to move
d. The force that keeps an object at rest

8. **What is the coefficient of friction?**
a. A value that represents the amount of friction between two surfaces
b. A value that represents the weight of an object
c. A value that represents the speed of an object
d. A value that represents the acceleration of an object

Answers to Multiple Choice Questions

1. (c); 2. (c); 3. (a); 4. (c); 5. (c); 6. (a); 7. (b); 8. (a)

REFERENCES

1. Al-Hashemi, H. M. B., & Al-Amoudi, O. S. B., (2018). A review on the angle of repose of granular materials. Powder Technology, 330, 397–417.
2. Archard, J. F., (1957). Elastic deformation and the laws of friction. Proceedings of the Royal Society of London; Series A Mathematical and Physical Sciences, 243(1233), 190–205.
3. Ayininuola, G. M., Agbede, O. A., & Franklin, S. O., (2009). Influence of calcium sulphate on subsoil cohesion and angle of friction. J. Appl. Sci. Res., 5(3), 297–304.
4. Battilana, J., & Lee, M., (2014). Advancing research on hybrid organizing–Insights from the study of social enterprises. Academy of Management Annals, 8(1), 397–441.
5. Bernadó, P., & Blackledge, M., (2010). Proteins in dynamic equilibrium. Nature, 468(7327), 1046–1048.
6. Blau, P. J., (2001). The significance and use of the friction coefficient. Tribology International, 34(9), 585–591.
7. Boroudjerdi, H., Kim, Y. W., Naji, A., Netz, R. R., Schlagberger, X., & Serr, A., (2005). Statics and dynamics of strongly charged soft matter. Physics Reports, 416(3, 4), 129–199.
8. Bowden, F. P., (1952). Introduction to the discussion: The mechanism of friction. Proceedings of the Royal Society of London. Series A. Mathematical and Physical Sciences, 212(1111), 440–449.
9. Braun, O. M., & Naumovets, A. G., (2006). Nanotribology: Microscopic mechanisms of friction. Surface Science Reports, 60(6, 7), 79–158.
10. Carlson, J. M., (2006). Functional limitations from pain caused by repetitive loading on the skin: A review and discussion for practitioners, with new data for limiting friction loads. JPO: Journal of Prosthetics and Orthotics, 18(4), 93–103.
11. Caron, O., Faure, B., & Brenière, Y., (1997). Estimating the center of gravity of the body on the basis of the center of pressure in standing posture. Journal of Biomechanics, 30(11, 12), 1169–1171.
12. Colebrook, C. F., & White, C. M., (1937). Experiments with fluid friction in roughened pipes. Proceedings of the Royal Society of London. Series A-Mathematical and Physical Sciences, 161(906), 367–381.
13. Cook, E. H., Buehler, M. J., & Spakovszky, Z. S., (2013). Mechanism of friction in rotating carbon nanotube bearings. Journal of the Mechanics and Physics of Solids, 61(2), 652–673.

14. Coplan, J. A., (1989). Rotational motion of the knee: A comparison of normal and pronating subjects. Journal of Orthopaedic & Sports Physical Therapy, 10(9), 366–369.
15. Dickson, C., Torabi, M., & Karimi, N., (2016). First and second law analyses of nanofluid forced convection in a partially-filled porous channel–the effects of local thermal non-equilibrium and internal heat sources. Applied Thermal Engineering, 103, 459–480.
16. Drummond, P. D., McNeil, K. J., & Walls, D. F., (1980). Non-equilibrium transitions in sub/second harmonic generation. Optica Acta: International Journal of Optics, 27(3), 321–335.
17. Eremeyev, V. A., & Altenbach, H., (2014). Equilibrium of a second-gradient fluid and an elastic solid with surface stresses. Meccanica, 49, 2635–2643.
18. Eyster, E., & Rabin, M., (2005). Cursed equilibrium. Econometrica, 73(5), 1623–1672.
19. Fang, H., & Parreiras, S. O., (2002). Equilibrium of affiliated value second price auctions with financially constrained bidders: The two-bidder case. Games and Economic Behavior, 39(2), 215–236.
20. Fischer, K. A., & Wriggers, P., (2006). Mortar based frictional contact formulation for higher order interpolations using the moving friction cone. Computer Methods in Applied Mechanics and Engineering, 195(37–40), 5020–5036.
21. Fusco, N., & Morini, M., (2012). Equilibrium configurations of epitaxially strained elastic films: Second order minimality conditions and qualitative properties of solutions. Archive for Rational Mechanics and Analysis, 203, 247–327.
22. Grether, J. M., & Mathys, N. A., (2010). Is the world's economic center of gravity already in Asia? Area, 42(1), 47–50.
23. Horstmann, G. A., & Dietz, V., (1990). A basic posture control mechanism: The stabilization of the center of gravity. Electroencephalography and Clinical Neurophysiology, 76(2), 165–176.
24. Jeon, B. G., Chang, S. J., Kim, S. W., & Kim, N. S., (2015). Base isolation performance of a cone-type friction pendulum bearing system. Struct. Eng. Mech., 53(2), 227–248.
25. Johannes, V. I., Green, M. A., & Brockley, C. A., (1973). The role of the rate of application of the tangential force in determining the static friction coefficient. Wear, 24(3), 381–385.
26. Kim, J., Kim, S., Schaumburg, E., & Sims, C. A., (2008). Calculating and using second-order accurate solutions of discrete time dynamic equilibrium models. Journal of Economic Dynamics and Control, 32(11), 3397–3414.
27. Kinoshita, M., (2009). Roles of translational motion of water molecules in sustaining life. Frontiers in Bioscience-Landmark, 14(9), 3419–3454.
28. Kruskal, M. D., & Oberman, C. R., (1958). On the stability of plasma in static equilibrium. The Physics of Fluids, 1(4), 275–280.
29. Lakhani, B., & Mansfield, A., (2015). Visual feedback of the center of gravity to optimize standing balance. Gait & Posture, 41(2), 499–503.

30. Lawton, D. T., (1983). Processing translational motion sequences. Computer Vision, Graphics, and Image Processing, 22(1), 116–144.
31. Lerner, A. P., (1937). Statics and dynamics in socialist economics. The Economic Journal, 47(186), 253–270.
32. Lincoln, B., (1953). Elastic deformation and the laws of friction. Nature, 172, 169–170.
33. Litao, C., Changyu, S., Guangjin, C., Yunqiang, N., Zhansong, S., & Yantao, L., (2009). Measurements of hydrate equilibrium conditions for CH_4, CO_2, and CH_4 + C_2H_6 + C_3H_8 in various systems by step-heating method. Chinese Journal of Chemical Engineering, 17(4), 635–641.
34. Lopez, G. E., Gurin, P., & Nagda, B. A., (1998). Education and understanding structural causes for group inequalities. Political Psychology, 19(2), 305–329.
35. Manow, P., Schäfer, A., & Zorn, H., (2008). Europe's party-political center of gravity, 1957–2003. Journal of European Public Policy, 15(1), 20–39.
36. Martins, J. A. C., Oden, J. T., & Simoes, F. M. F., (1990). A study of static and kinetic friction. International Journal of Engineering Science, 28(1), 29–92.
37. McNelley, T. R., Swaminathan, S., & Su, J. Q., (2008). Recrystallization mechanisms during friction stir welding/processing of aluminum alloys. Scripta Materialia, 58(5), 349–354.
38. Medlin, B., & Green, K. W., (2009). Enhancing performance through goal setting, engagement, and optimism. Industrial Management & Data Systems, 109(7), 943–956.
39. Melander, W. R., Lin, H. J., Jacobson, J., & Horvath, C., (1984). Dynamic effect of secondary equilibria in reversed-phase chromatography. The Journal of Physical Chemistry, 88(20), 4527–4536.
40. Miura, K., Maeda, K., & Toki, S., (1997). Method of measurement for the angle of repose of sands. Soils and Foundations, 37(2), 89–96.
41. Mo, Y., Turner, K. T., & Szlufarska, I., (2009). Friction laws at the nanoscale. Nature, 457(7233), 1116–1119.
42. Morgul, O., (1991). Orientation and stabilization of a flexible beam attached to a rigid body: Planar motion. IEEE Transactions on Automatic Control, 36(8), 953–962.
43. Orlov, A., (2015). An assessment of proposed energy resource tax reform in Russia: A static general equilibrium analysis. Energy Economics, 50, 251–263.
44. Parker, J. C., & Valocchi, A. J., (1986). Constraints on the validity of equilibrium and first-order kinetic transport models in structured soils. Water Resources Research, 22(3), 399–407.
45. Partridge, M. D., Rickman, D. S., Olfert, M. R., & Tan, Y., (2015). When spatial equilibrium fails: Is place-based policy second best? Regional Studies, 49(8), 1303–1325.
46. Popova, E., & Popov, V. L., (2015). The research works of Coulomb and Amontons and generalized laws of friction. Friction, 3, 183–190.
47. Ramos, A. G., Oliveira, J. F., Gonçalves, J. F., & Lopes, M. P., (2016). A container loading algorithm with static mechanical equilibrium stability constraints. Transportation Research

Part B: Methodological, 91, 565–581.

48. Raudenbush, S. W., (2008). Advancing educational policy by advancing research on instruction. American Educational Research Journal, 45(1), 206–230.
49. Rossouw, P. E., (2003). Friction: An overview. In: Seminars in Orthodontics (Vol. 9, No. 4, pp. 218–222). Elsevier.
50. Schanz, T., & Vermeer, P. A., (1996). Angles of friction and dilatancy of sand. Géotechnique, 46(1), 145–151.
51. Schlanbusch, R., Loria, A., & Nicklasson, P. J., (2012). On the stability and stabilization of quaternion equilibria of rigid bodies. Automatica, 48(12), 3135–3141.
52. Schuler, A. R., Grammatikos, A., & Fegley, K. A., (1967). Measuring rotational motion with linear accelerometers. IEEE Transactions on Aerospace and Electronic Systems, (3), 465–472.
53. Shaw, S. W., (1986). On the dynamic response of a system with dry friction. Journal of Sound and Vibration, 108(2), 305–325.
54. Shinjo, K., & Hirano, M., (1993). Dynamics of friction: Superlubric state. Surface Science, 283(1–3), 473–478.
55. Smith, W. K., & Lewis, M. W., (2011). Toward a theory of paradox: A dynamic equilibrium model of organizing. Academy of Management Review, 36(2), 381–403.
56. Steif, P. S., & Dollar, A., (2005). Reinventing the teaching of statics. International Journal of Engineering Education, 21(4), 723.
57. Stiny, G., & March, L., (1981). Design machines. Environment and Planning B: Planning and Design, 8(3), 245–255.
58. Suh, N. P., Mosleh, M., & Howard, P. S., (1994). Control of friction. Wear, 175(1, 2), 151–158.
59. Takashi, N., & Hughes, T. J., (1992). An arbitrary Lagrangian-Eulerian finite element method for interaction of fluid and a rigid body. Computer Methods in Applied Mechanics and Engineering, 95(1), 115–138.
60. Tewary, V. K., (1973). Green-function method for lattice statics. Advances in Physics, 22(6), 757–810.
61. Urbakh, M., Klafter, J., Gourdon, D., & Israelachvili, J., (2004). The nonlinear nature of friction. Nature, 430(6999), 525–528.
62. Van, B. A., (1945). Angle of repose and angle of sliding friction: An experimental study. Geological Society of America Bulletin, 56(6), 669–707.
63. Wachter, J. K., & Yorio, P. L., (2014). A system of safety management practices and worker engagement for reducing and preventing accidents: An empirical and theoretical investigation. Accident Analysis & Prevention, 68, 117–130.
64. Wang, Y., & Xu, S., (2013). Stability of the classical type of relative equilibria of a rigid body in the J 2 problem. Astrophysics and Space Science, 346, 443–461.
65. Watling, D., (2002). A second order stochastic network equilibrium model, I: Theoretical foundation. Transportation Science, 36(2), 149–166.

66. Zagare, F. C., (1984). Limited-move equilibria in 2× 2 games. Theory and Decision, 16, 1–19.
67. Zhang, H. M., & Ge, Y. E., (2004). Modeling variable demand equilibrium under second-best road pricing. Transportation Research Part B: Methodological, 38(8), 733–749.

CHAPTER 4

MECHANICS OF MATERIALS

UNIT INTRODUCTION

A fundamental area of physics and engineering is the study of materials mechanics, which examines how solid materials respond to stresses or loads from the outside world. To comprehend the behavioral patterns and design of different engineering elements and structures, like beams, shafts, columns, and plates, it is crucial to understand the basic principles and ideas of material mechanics. These principles and ideas are summarized in this chapter (Muhanna et al., 2007). To determine the strains, internal stresses, and solid deformations exposed to external loads, the mechanics of materials apply the mechanics' principles, such as dynamics, statics, and strength of materials. The chapter provides examples to explain the principles and ideas, covering areas including tensile test, stress, and strain analysis, shear, compound bars, and bending (Daniel et al., 2006). The science of materials mechanics is essential for their construction and optimization to assure the safety, dependability, durability, and effectiveness of different engineering systems, like bridges, buildings, automobiles, aircraft, and machinery (Gross et al., 2011).

Learning Objectives

At the end of this chapter, readers will be able to:

- Understand the concepts of stress and strain and their relationship.

- Familiarize with the tensile test and its importance in determining mechanical properties of materials.
- Differentiate between brittle and ductile materials.
- Define and calculate Poisson's ratio for materials.
- Understand the concepts of shear stress and strain and their relationship.
- Calculate and analyze stress and strain in compound bars.
- Understand and analyze external loads exerted on compound bars.
- Analyze the effects of temperature fluctuations on compound bars.
- Analyze and draw bending moment and shearing force diagrams.
- Understand and apply bending theory to determine stresses and strains in beams.
- Calculate and determine the neutral axis of beams.
- Understand the concept and application of built-in beams.
- Evaluate the pros and cons of built-in beams.
- Differentiate between different types of instability.

Key Terms

- Bending moment
- Bending theory
- Brittle materials
- Built-in beam
- Compound bars
- Ductile materials
- Instability
- Neutral axis
- Poisson's ratio
- Shear stress and strain
- Shearing force diagrams
- Strain
- Stress
- Tensile test

4.1. STRESS AND STRAIN

Simple stress and strain are basic principles in material mechanics that describe how solid materials behave when subjected to loads or forces from the outside world. When a material is exposed to external pressures like compression, bending, tension, or shear, an internal force per unit area known as stress occurs. The force divided by the area of cross-sectional of the material to determine stress, is measured in unit pascals (Pa) also pounds per square inch (psi). A material's geometry, loading circumstances, and material characteristics, like plasticity and elasticity, all affect the amount and dispersion of stress in that material (Hartsuijker & Welleman, 2007).

Remember

Stress is the force per unit area on a material, and strain is the resulting deformation per unit length. They are related by the modulus of elasticity and are important in designing materials and structures.

When a substance is stressed, it deforms or changes shape, referred to as a strain. Strain is a quantity with no dimensions determined as the difference between the dimension or actual length of the material divided by the new length or dimension. The elastic modulus of the object, a gage of its resistance and stiffness to deformation, connects strain to stress. The curve of stress-strain, which depicts the response of the material under different loading situations and can be utilized to ascertain its mechanical characteristics, like yield ultimate strength, strength, and elasticity modulus, describes the stress-strain interaction (Kubıcek & Florian, 2009) (Figure 4.1).

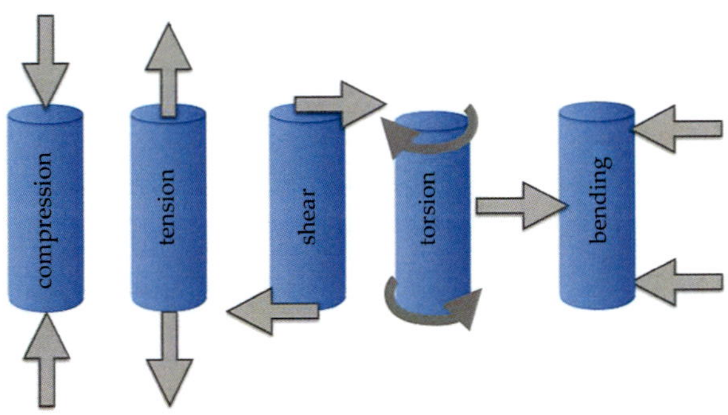

Figure 4.1. Illustration of stress and strain.

Source: Linear Motion Tips, Creative Commons License.

4.1.1. Load

The parts of any engineering system will be vulnerable to external forces resulting from the service parameters in which the part operates. The resultant will be zero for the external forces if the component is in equilibrium. However, they still exert a load likely to deform it and to which internal forces built within the substance must respond. A cylindrical bar is considered under tension or compression if it experiences a direct push or pulls along its axis. While compression happens in the chair's legs as you sit in it, tension is typically represented by the forces in pulling ropes equipment (Fischer et al., 2014).

Although one newton is a fairly less load in engineering terms, the SI system measures loads in Newton. As a result, loads are typically expressed in SI multiples like meganewtons or kilonewtons in engineering applications (MN).

The application of load to a member could be made in various ways. Common loading types include:

- Gravity effects generally cause static and non-fluctuating loads.
- Impact loads produced by a blow.
- Fluctuating, fatigue, the sign and magnitude of the load fluctuating with respect to time.
- Liue loads, as exerted by traffic passing over a bridge (Jiao & Moan, 1990).

4.1.2. Normal or Direct Stress (a)

As was said earlier, any external force acting on an equilibrium body will cause internal forces to be created within the substance. So, if a bar is confined to uniform compression or tension that is applied evenly or uniformly over the cross-section, the internal forces created are also constantly spread, and the bar is considered to be exposed to constant normal stress. It is described as (Phillips & Heywood, 1951):

$$\text{stress } (\sigma) = \frac{\text{load}}{\text{area}} = \frac{P}{A}$$

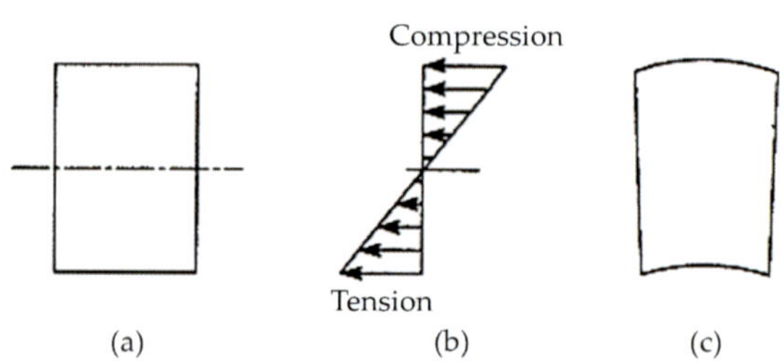

Figure 4.2. Diagram of direct stress distribution.

Source: Direct Science, Creative Commons License.

Hence, stress CT would be calculated in newtons per square meter (N/mZ) and may be tensile or compressive based on the type of load. The stress at certain location is determined by the limiting factor of 6P/6A since 6A tends to zero in various loading situations where the stress will differ over any particular section (Zhu & Huang, 2019) (Figure 4.2).

Direct strain (E): A bar will alter in length if put under a direct load, creating stress. The strain created is discussed if the bar has an initial length of L and lengthens by a factor of 6:

$$\text{strain}\,(\varepsilon) = \frac{\text{change in length}}{\text{orginal length}} = \frac{\delta L}{L}$$

Hence, strain is a ratio of two values with the same unit that measures the material's deformation and is dimensionless.

Since, in practice, the extensions of materials under load are very small, it is often convenient to measure the strains in the form of strain x 10^{-6} i.e. microstrain, when the symbol used becomes $\mu\varepsilon$

Alternatively, strain can be expressed as a percentage strain (Neggers et al., 2014).

$$\text{strain}\,(\varepsilon) = \frac{\delta L}{L} \times 100\%$$

4.1.3. Hooke's Law – Elastic Materials

When a load is released, a material is considered elastic if its initial, unloaded dimensions are restored. A certain type of elasticity causes deformations proportionate to the loads causing them in a wide variety of materials, at least across a portion of their load range. This also indicates that, even though substances are elastic, stress is proportional to strain since weights are proportional to the stresses they induce, and distortions are related to the strains (Dell'Isola et al., 2009). Hence, Hooke's law, in its most basic version, asserts that a strain (E) and stress (a)

$$\text{stress}\,(\sigma) \propto \text{strain}\,(\varepsilon)$$

$$\frac{\text{stress}}{\text{strain}} = \text{constant*}$$

As demonstrated in the following sections, most ferrous alloys adhere to this law within specific bounds. It might be reasonably expected to extend to other engineering materials like wood, non-ferrous alloys, and concrete. The deformation caused by any load while the elastic material will be sustained when the load is removed; there is no lasting distortion. The following are some categories of information that the reader must be familiar with (Nordmann et al., 2020) (Figure 4.3).

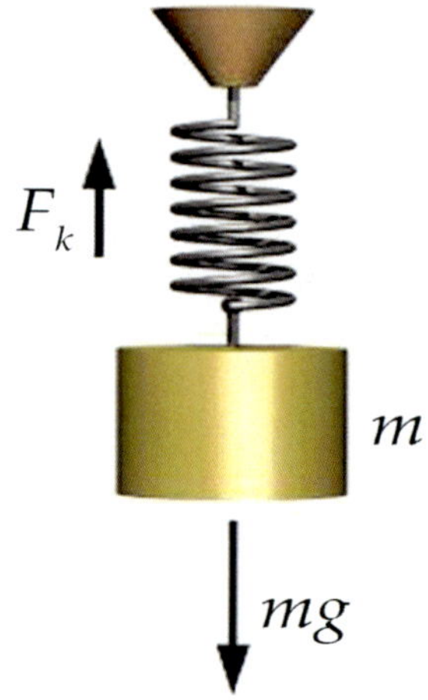

Figure 4.3. Schematic of Hook's law.

Source: D. Klepper, Creative Commons License.

A homogeneous material is precisely consistent and free of defects or discontinuities. Inhomogeneous materials or n on-homogeneous will therefore possess a structure that changes at any point based on the elements and the presence of impurities or casting defects. Examples of these materials include concrete and low-quality iron. A substance is described as isotropic if its qualities are uniform in all dimensions; conversely, a substance that does not reflect this uniform behavior is described as anisotropic or noninotropic (Yang et al., 1998).

A substance that differs in characteristics in various planes is considered orthotropic. Wood is a common example of a material in this domain, while other composites that include systematically oriented "inhomogeneities" can also be included (Rychlewski, 1984).

4.1.4. Young's Modulus

It has been demonstrated that within the elastic constraints of materials or within the ranges in which Hooke's law applies.

$$\frac{\text{stress}}{\text{strain}} = \text{constant}$$

here; E is the constant and known as the Young's modulus or elasticity modulus.

Thus,

$$E = \frac{\text{stress}}{\text{strain}} = \frac{\sigma}{\varepsilon} \quad (4.1)$$

Young's modulus E, which has a large numerical value for most engineering materials, is typically believed to be the same in compression or tension. For steel, E = 200 × 10^9 N/m^2, hence it can be seen from (1.1) that strains are often quite small, since

$$= \frac{P}{A} \div \frac{\delta L}{L} = \frac{PL}{A\delta L} \qquad (4.2)$$

$$\propto \text{strain}\,(\varepsilon)$$

$$\varepsilon = \frac{\sigma}{E}$$

The hypothesis that deformations are minor compared to original dimensions, which is utilized later in the text, is typically well-founded because strains rarely surpass 0.003 or 0.3% in most conventional engineering applications. Any material's actual Young's modulus value is typically ascertained by performing a conventional tensile test on a material sample (Yamaguchi et al., 1998) (Figure 4.4).

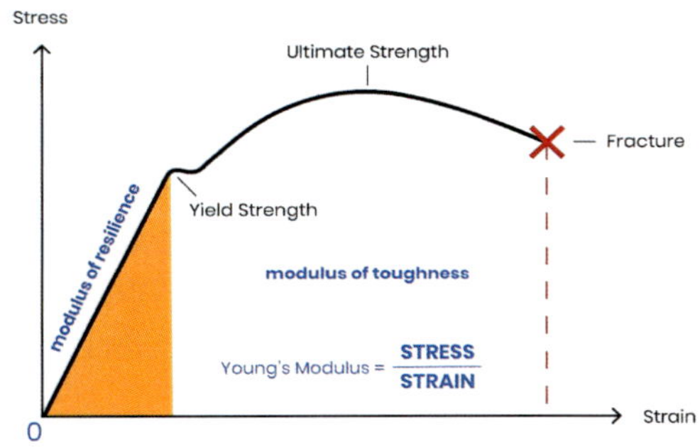

Figure 4.4. Graphical representation of Young's modulus.

Source: Aramid, Creative Commons License.

4.2. TENSILE TEST

It is required to perform some standardized experiments to measure the relative qualities of the materials in order to contrast the strengths of different materials. One test is the conventional tensile test, which involves applying a circular bar with a homogeneous cross-section to an escalating tensile load until collapse. Extensometers measure the bar's length as it changes during the loading process. The results are plotted as a graph of load for stress against strain. Other materials will represent distinct graphs, but they will all follow a similar general format (Hart, 1967) (Figure 4.5).

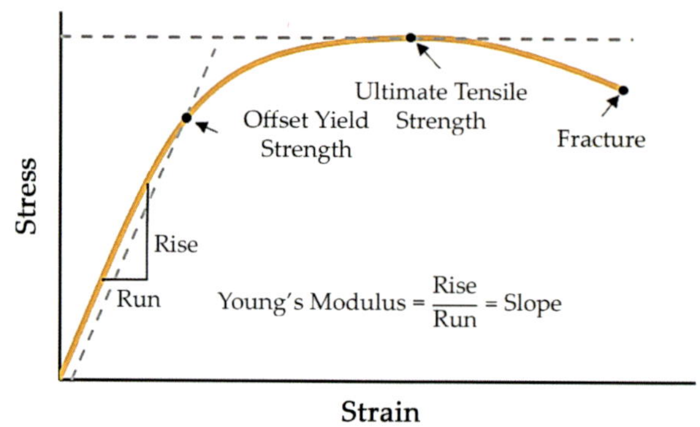

Figure 4.5. Graph of tensile testing.

Source: Standard Mechanics, Creative Commons License.

For the initial phase of the experiment, it would be seen that Hooke's law is upheld, meaning that the material acts elastically, and that stress and strain are proportionate, as shown by the graph. The graph finally reaches point A, known as the limit of proportionality, where the graph's linearity ends. After this point, the material may continue to be elastic with respect to fully recovered deformations once the load is eliminated (i.e., strain falls to zero), though Hooke's law is not applicable. The elastic limit for this situation is limiting point B. Points A and B are frequently taken to be identical for all reasons. Plastic deformation occurs beyond the elastic limit, and strains cannot fully recover (Davis, 2004). As a result, when the load is released, there will be a certain permanent deformation. After the so-called upper yield point (point C) and the lower yield point (point D), relatively quick raises in strain happen without matching significant rise in stress. As a result, the graph gets significantly shallower and spans a lot larger area of the strain axis than the material's elastic range does. A material's so-called flexibility, which will be covered in more detail below, is measured by its ability to withstand these significant plastic deformations (Cabezas & Celentano, 2004).

Remember

Leonardo da Vinci conducted the first recorded tensile test in the late 15th century, using a system of weights and pulleys to measure the strength of wires made from various metals. This paved the way for the modern version of the tensile test used in materials science and engineering.

It is impossible to distinguish between the lower and upper yield points for some materials, such as non-ferrous metals and high-carbon steels, and in some circumstances, there is no yield point in any way. In these situations, proof stress is employed to signal the beginning of plastic strain or to compare the relative qualities with those of another similar material. The persistent deformation caused by a loading cycle is measured here; for instance, the 0.1% proof stress is the stress that, when withdrawn, results in a permanent strain or "set" of 0.1% of the initial gage length (Figure 4.6(a)). The following query (Tvergaard et al., 1981):

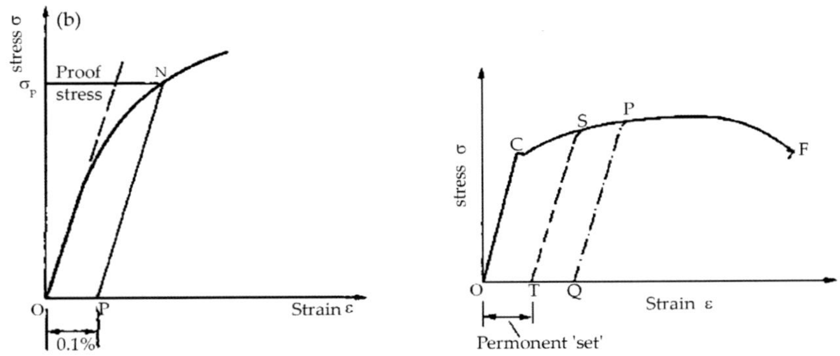

Source: E. J. Hearn, Creative Commons License.

Figure 4.6. (a) Determination of 0.1% proof stress; and (b) permanent deformation or "set" after straining beyond the yield point.

Note the 0.1% strain equivalent at point P on the strain axis. To break the curve in N, start at P and draw a line parallel to the tensile test curve's straight-line section. The N stress follows the 0.1% proof stress. If the permanent set is less than 0.1% after the proof stress has been exerted for 15 seconds and withdrawn, the material is said to meet its specifications. To apply strain past the yield point and to point E on the graph, there must be some increase in load. The material is in an elastic-plastic condition between points D and E, with some sections staying elastic and helping to restore the original dimensions when the load is lifted, and the remainder must be plastic (LaVan & Sharpe, 1999). Beyond E, the bar is termed to neck when its cross-sectional area starts to decrease quickly over a fairly short length of the bar. When the stress decreases, the bar begins to neck, and at point F, breakage eventually takes place. The load at point E divided by the bar's initial cross-sectional area results in the nominal stress at the point of failure, also called the ultimate tensile stress (the tensile strength of the bar's material is another name for this). The stress at breakage is frequently higher than the number mentioned above because of the significant area reduction caused by the necking action. Moreover, as designers are more concerned with the maximum loads the entire cross-section can support, the stress at breakage rarely has real-world significance (Hannon & Tiernan, 2008).

According to Figure 4.6(b), the offloading line ST can be assumed to be straight for the majority of practical reasons if the load is lifted from the test sample after the yield point C has been crossed, for example, to some position S. Hence, the offloading mechanism is entirely elastic even though the loading to S has both plastic (CS) and elastic (OC) components. Following the line TS and continuing along the initial curve to break down at F, a second load cycle would begin with the permanent extension linked to the strain OT. Yet, it will be shown that the dynamic load cycle expands the material's elastic range, moving the actual yield point from C to S yet leaving the tensile strength unaffected. The material is deemed work-hardened if the process was continued along the lines of PQ, etc. (Gianola & Eberl, 2009).

Because the offloading and reloading lines will create a small loop and are no longer perfectly linear, thorough observation reveals that the materials would no longer reflect real elasticity. While the extension or strain to breakdown will be significantly reduced, continuous offloading and reloading will result in a yield point close to the final stress value (Lord & Morrell, 2010). In Figures 4.7–4.9, standard curves from tensile testing on other technical materials are depicted.

Figure 4.7. Illustration of tensile test curves for various metals.

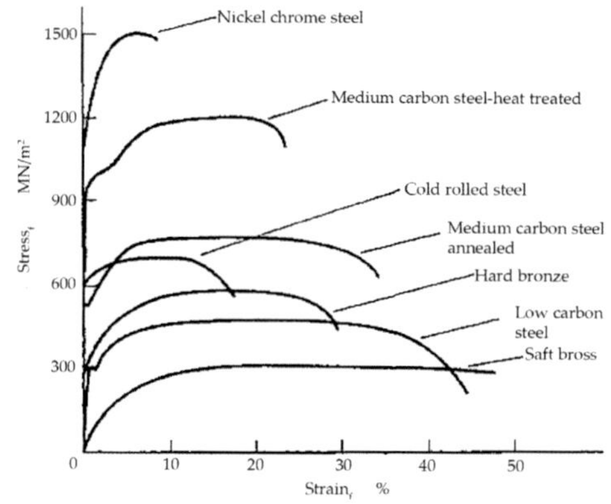

Source: E. J. Hearn, Creative Commons License.

Figure 4.8. Graphical representation of typical stress-strain curves for hard drawn wire materials.

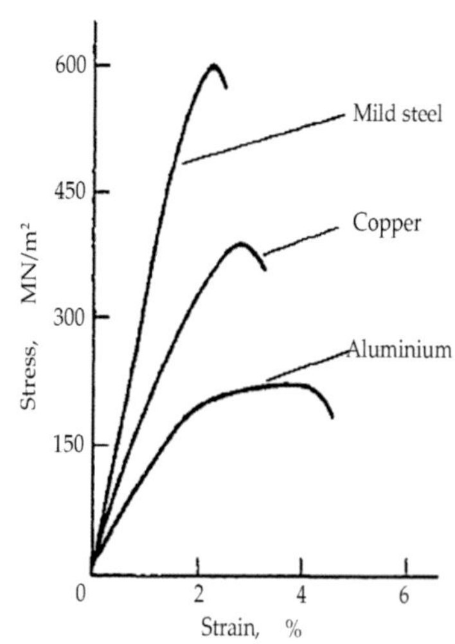

Source: University of Warwick, Creative Commons License.

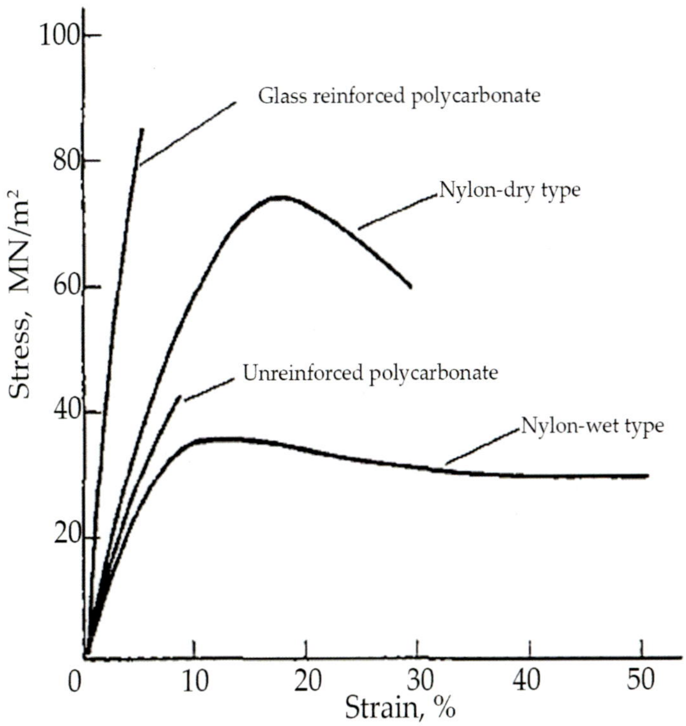

Figure 4.9. Graph of typical tensile test results for various types of nylon and polycarbonate.

Source: University of Warwick Creative Commons License.

It is typically essential to consult a "Code of Practice" or "British Standard Specification" after doing the standard tensile test to make sure the material tested complies with the specifications:

- BS 970 British Standard specification for wrought steels.
- BS 4360 British Standard specification for weldable structural steels.
- BS 153 British Standard specification for steel girder bridges.
- BS 449 British Standard specification for using structural steel in building (Taylor & Elam, 1923).

4.3. BRITTLE AND DUCTILE MATERIALS

The graph of somewhat plastic range in Figure 4.10 spans a substantially larger section of the strain axis than the elastic range, as was already seen. As a result, the material's elongation over this range far exceeds that caused by elastic loading. The ability of a material to permit such substantial expansions, or to be pulled out plastically, is referred to as ductility. High ductility materials are known as ductile materials, and low ductility materials are known as brittle materials. By making calculation of the percentage reduction or extension in area, both given below, one can determine the quantitative measure of ductility (Aquaro & Fontani, 2001) (Figure 4.10).

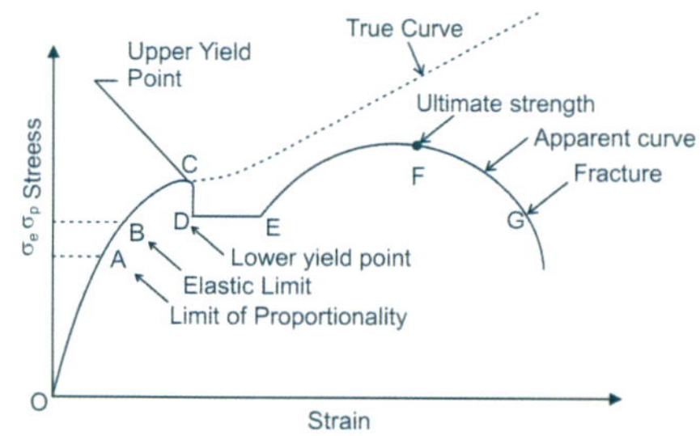

Figure 4.10. Diagram of typical tensile test curve for mild steel.

Source: Test Book, Creative Commons License.

$$\text{Percentage elongation} = \frac{\text{increase in guage legth to fracture}}{\text{original guage length}} \times 100$$

$$\text{Percentage reduction in area} = \frac{\text{reduction in cross} - \text{sectional area of necked portion}}{\text{original area}} \times 100$$

The latter number is typically considered the more relevant measure of ductility for comparison reasons because it is irrespective of any specified gage length.

The capacity of a material to be pounded into thin sheets is known as malleability, a quality closely connected to ductility. Lead is a common illustration of a malleable material. This is frequently employed in plumbing, hammered into edges or joints to create a waterproof seal. Hence, malleability is defined as a material's capacity to permit continuous lateral extensions under compressive loads (Masoero et al., 2010).

A brittle material has small extensions to breakage, significantly reducing the largely plastic section of the tensile test graph (Figure 4.11).

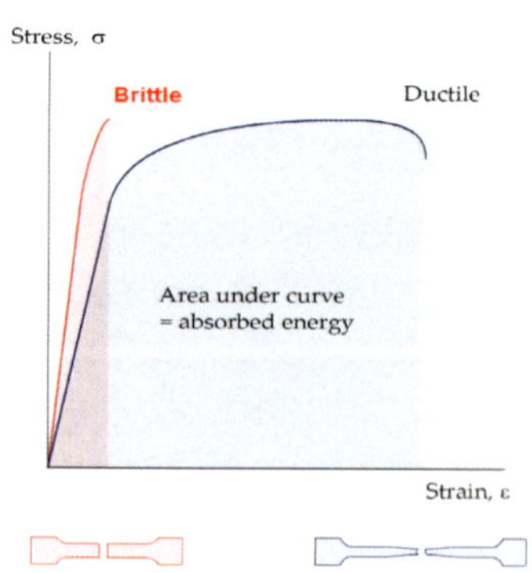

Figure 4.11. Graph of typical tensile test curve for a brittle material.

Source: B.Tech, Creative Commons License)

4.4. POISSON'S RATIO

Suppose a tensile load is applied to the rectangular bar in Figure 4.12. The bar will lengthen by 6 L under the influence of this weight, causing a longitudinal strain in the bar of

$$\varepsilon_L = \frac{\delta L}{L}$$

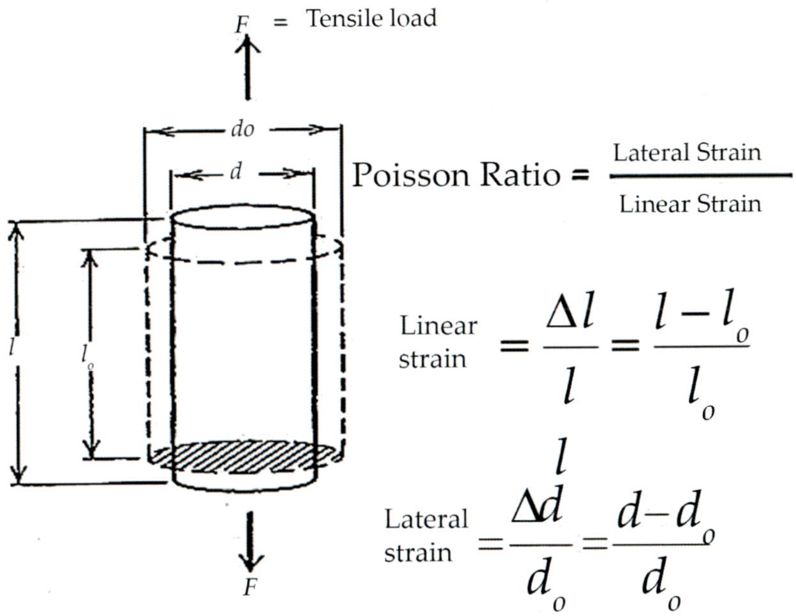

Figure 4.12. Illustration of Poisson's ratio.

Source: What is Piping, Creative Commons License.

Yet the bar will also show a lateral dimension reduction, meaning that its width and depth will be reduced (Wingender & Balzani, 2022). The lateral strains that go along with it will both be equal, have the opposite meaning as the longitudinal strain, and be determined by:

$$\varepsilon_{lat} = -\frac{\delta b}{b} = -\frac{\delta d}{d}$$

The longitudinal and lateral strains' ratio will remain unchanged as long as the material load is kept within the elastic range. Poisson's ratio is the name for this ratio (Li et al., 2021).

$$\text{Poisson's ratio } (v) = \frac{\text{lateral strain}}{\text{longitudinal strain}} = \frac{(-\delta d/d)}{\delta L/L} \qquad (4.3)$$

Normal practice ignores the lateral strain's negative sign, leaving Poisson's ratio as merely a strain magnitude ratio. Moreover, it should be offered that a longitudinal strain causes a lateral strain of the opposite sign; for example, a lateral compressive strain causes a tensile longitudinal strain.

The value of v for most engineering materials ranges from 0.25 to 0.33 (Lee & Lakes, 1997). As:

$$\text{longitudinal strain} = \frac{\text{longitudinal stress}}{\text{Young's modulus}} = \frac{\sigma}{E}$$

$$\text{lateral strain} = \frac{\sigma}{E} \qquad (4.4)$$

4.5. SHEAR STRESS AND STRAIN

Assume a block or piece of material in Figure 4.13(a) is being pulled in a certain direction by a system of equal and opposing forces Q. (One such mechanism might be implemented in the brake block of a bicycle when it makes contact with the wheel.) The material, therefore, has the propensity to slide one layer over another, leading to the type of breakdown depicted in Figure 4.13(b). If this breakdown is constrained, the following shear stress T is created (Sabik, 2018):

$$\text{shear stress}(\tau) = \frac{\text{shear load}}{\text{area resisting shear}} = \frac{Q}{A}$$

Direct stresses are always typical to the region on which they affect. But this shear stress, therefore, would be tangential to the surface on which it affects.

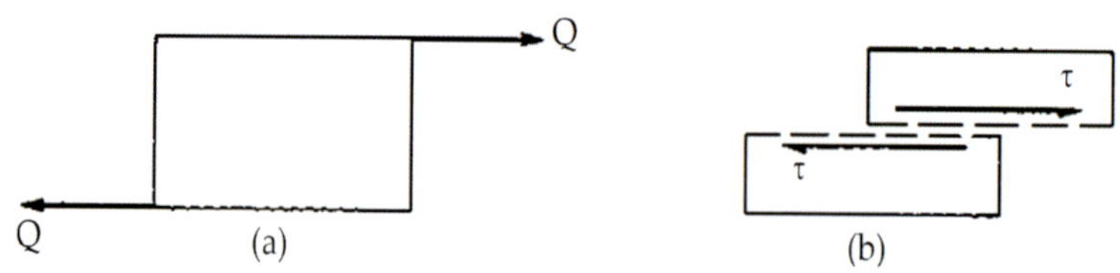

Figure 4.13. Shear force and resulting shear stress system showing typical form of failure by relative sliding of planes.

Source: E. J. Hearn, Creative Commons License.

Shear Strain: The block's rectangular shape would not be sustained once the brake is pressed, and the shear forces are produced if the block of Figure 4.13(a) is once more considered a bicycle brake block. The block will, in fact, deform or "strain" into the shape as shown below. The shear strain is then used to describe the deformation angle y.

Shear strain is dimensionless, or without units, because it is calculated in radians (McGarry et al., 2005).

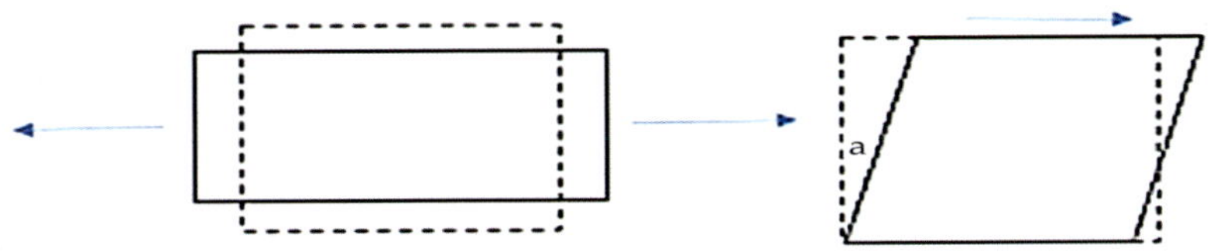

Normal Strain Shear Strain

4.6. COMPOUND BARS

When a compound bar made of various lengths, widths, and materials is put under compressive load W or external tensile, the load borne by every member is represented by:

$$F_1 = \frac{\dfrac{E_1 A_1}{L_1}}{\dfrac{EA}{L}} W$$

When a member is mentioned by the suffix 1, the aggregate of all such quantities for all members is indicated.

The compound bar could be broken down into a single comparable bar with a similar Young's modulus, known as a combined E, when the bars share a length (Chen & Weng, 2001).

$$\text{Combined } E \quad \frac{\Sigma EA}{\Sigma A}$$

Through a temperature shift from T_1 to T_2, a bar's free expansion is given as:

$$\alpha(T1 - T2)L'$$

where; L denotes the bar length; and a is the coefficient of linear expansion. If the expansion is inhibited, Stress would be generated in the bar represented by:

$$\alpha(T1 - T2)E$$

Two principles are applied to measure the stresses in a compound bar made up of two members with varying lengths:

- The compressive force exerted on the long member by the short member is proportional to the tensile force exerted on the short member by the long member.
- The variation in free lengths is equal to the short member's extension plus the long member's compression (Bert & Zeng, 2001).

TIPS

Design compound bars with materials of higher coefficients of thermal expansion on the outer edges to minimize stress and deformation from temperature changes. Consider material properties and temperature range when designing compound bars.

When two components of different materials—with differing expansion coefficients—that were initially equal in length experience a temperature change or tighten a nut, the free lengths may differ. If this bar type is subsequently exposed to a further external load, the resultant stresses may be calculated with superposition concept. The resultant stresses are generated using this method by adding the results while accounting for a sign to derive the stresses in the members resulting from the individual effects (Senitsky & Ishutin, 2015).

4.6.1. External Load Exerted on Compound Bars

In some applications, combining pieces or bars composed of various materials may be required, each of which serves a different purpose. For instance, it is commonly practicable to encounter the current in a system of copper wires encircling steel wires in overhead electrical cords because the latter is designed to support the stress of the cord over long distances. The common term for any such material configurations is compound burs (Saadatmanesh et al., 1994).

Such a compound bar distributes external stress among its constituent materials in proportions based on their particular areas, Young's moduli, and lengths.

So, suppose a compound bar of n parts, each with a distinct cross-sectional area, length, and material. Assume that the load is placed to cause the same expansion in each member, giving all members a common extension of x (Emri et al., 2016).

For the nth member,

$$\frac{\text{stress}}{\text{strain}} = E_n = \frac{F_n L_n}{A_n x_n}$$

$$F_n = \frac{E_n A_n x_n}{L_n} \qquad (4.5)$$

A and Ln are the cross-sectional areas and length of the nth member, and F is the force acting on that member. The aggregate of these loads for all members will represent the complete load kept.

$$W = \sum \frac{E_n A_n X}{L_n} = x \sum \frac{E_n A_n}{L_n} \quad (4.6)$$

Now from Eqn. (4.5) the force in member 1 is given by:

$$F_1 = \frac{E_1 A_1 X}{L_1} \quad (4.6)$$

But, from Eqn. (4.6)

$$x = \frac{W}{\sum \frac{E_n A_n}{L_n}}$$

$$F1 = \frac{\frac{E_1 A_1}{L_1}}{\sum \frac{EA}{L}} W \quad 4.7$$

According to its EAIL value, each member bears a percentage of the overall load W. The given equation simplifies if all of the wires are the same length.

$$F1 = \frac{E_1 A_1}{\sum EA} W \quad (4.8)$$

The stress in member 1 is written as

$$\sigma_1 = \frac{F_1}{A_1} \quad (4.9)$$

"Combined" modulus or Compound bars – "equivalent."

It is customary to think of a compound bar as a single bar made of a hypothetical material with a combined modulus E in order to calculate its common extension. Thus, it is crucial to assume that each component of the compound bar has the same extension and initial lengths since only then would all of the components' strains be equal.

Now complete load on compound bar = $F_1 + F_2 + F_3 + \ldots + F$, where F_1, F_2, etc., are the loads in members 1, 2, etc. (Okelo, 2002).

But

Force=Stress*area

$$\therefore \quad \sigma(A_1 + A_2 + ... + A_n) = \sigma_1 A_1 + \sigma_2 A_2 + ...\sigma_n A_n$$

Where **α** is the stress in the equivalent single bar.

Dividing through by the common strain

$$\frac{\sigma}{\varepsilon}(A_1 + A_2 + ... + A_n) = \frac{\sigma_1}{\varepsilon}A_1 + \frac{\sigma_2}{\varepsilon}A_2 + ...\frac{\sigma_n}{\varepsilon}A_n$$

$$E_c(A_1 + A_2 + ... + A_n) = E_1 A_1 + E_2 A_2 + ...E_n A_n$$

Where E_C is the equivalent or combined E of the single bar

$$\therefore \text{ combined } E = \frac{E_1 A_1 + E_2 A_2 + ...E_n A_n}{A_1 + A_2 + ...A_n}$$

$$\text{i.e.} \quad E_c = \frac{\sum EA}{A}$$

With an external load W applied

$$\therefore \text{ combined } E = \frac{E_1 A_1 + E_2 A_2 + ...E_n A_n}{A_1 + A_2 + ...A_n}$$

$$\text{i.e.} \quad E_c = \frac{\sum EA}{A}$$

With an external load W applied,

$$\text{and} \quad \text{stress in the equivalent bar} = \frac{W}{\sum A}$$

$$\text{strain in the equivalent bar} = \frac{W}{E_c \sum A} = \frac{x}{L}$$

$$\therefore \text{since} \quad \frac{\text{stress}}{\text{strain}} = E$$

$$\text{common extension } x = \frac{WL}{E_c \sum A}$$

$$= \text{extention of sin gle bar}$$

4.6.2. Temperature Fluctuations on Compound Bars

When exposed to a temperature fluctuation, a material's length will fluctuate by a certain level.

ΔL=aLt

where; L denotes the initial length; t denotes the temperature fluctuation; and a denotes the material's linear expansion coefficient. (Unless in very unusual circumstances involving materials with negative or zero expansion coefficients, which should not be supposed here, a temperature rise results in an elongation in length, and a temperature drop generates a decrease in length.) Yet, stress is created within the substance if an outside force prevents the material from expanding freely. This stress is equivalent in size to the tension created in the bar if its length was originally allowed to alter freely before being forced back to its initial length (Duc et al., 2002).

$$\text{change in length} = \alpha L t$$

$$\text{strain} = \frac{\alpha L t}{L} = \alpha t$$

Thus, the tension that is produced in the material when enough force is applied to release this strain.

$$= \text{strain} \times E$$

$$= E\alpha t$$

Now consider a compound bar of two tightly linked materials, as shown in below. Assume that the materials in this situation are brass and steel for simplicity.

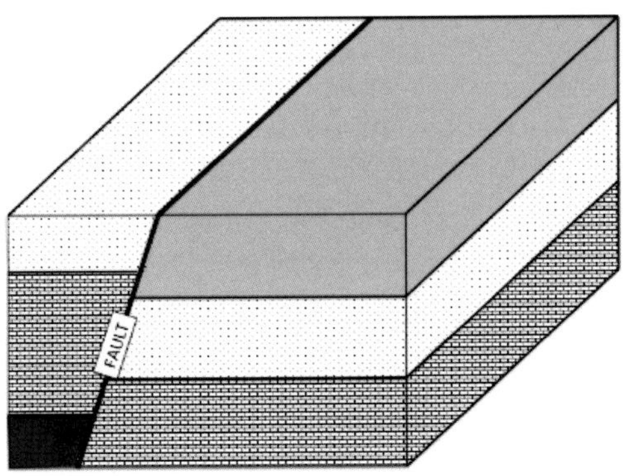

The two materials that make up the compound bar will typically have distinct coefficients of expansion, which means that when the temperature increases, every material will strive to stretch by a certain number. Figure 4.14(b) depicts the positions each material will stretch to if it is given complete freedom to expand (not connected firmly as a compound bar). Every expansion of length L is specified by:

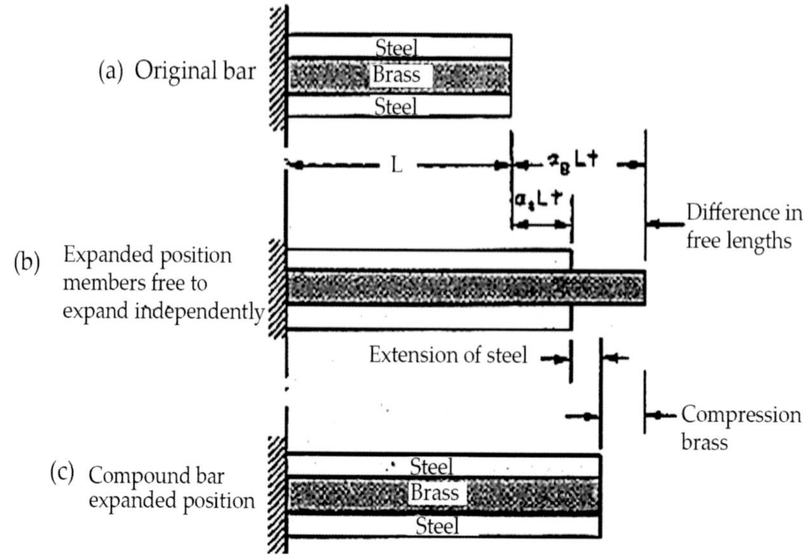

Figure 4.14. Diagram of thermal expansion of compound bar.

Source: University of Warwick, Creative Commons License.

Hence, the variation in "free" expansion or "free lengths" (Krishnan et al., 2018).

$$= \alpha_B Lt - \alpha_S Lt = (\alpha_B - \alpha_S)Lt$$

Since the expansion coefficient of the brass IXB in this instance is higher in comparison to steel IX. It is assumed to be the equal initial lengths L of the two materials. The two metals will strive to expand to their free-length positions if they are rigidly connected as a compound bar and heated to the same degree, but one will be influenced by the movements of the other. Hence, the steel's "free length" position will be sought after by the lower coefficient of expansion material (steel). In comparison, the brass's "free length" position will be sought after by the higher coefficient of expansion material (brass). A solution is made, and the compound bar extends to the position. This causes the steel to stretch from its free-length position and the brass to compress out of its free-length position effectively. The illustration will demonstrate that the given rule is valid (Alekseevskii et al., 1990).

Ø 1st Rule:

Compression of brass + Extension of steel = difference in "free" lengths.

Compression of "long" member + Extension of "short" member = difference in free lengths.

2nd Rule 2: The compressive force exerted on the short member by the short member is proportional to the tensile force exerted on the short member or by the long member.

Thus, in this case

Compressive force in brass = tensile force in steel

Now Young's modulus definition is given as

$$E = \frac{\text{stress}}{\text{strain}} = \frac{\sigma}{\delta / L}$$

Where δ is the change in length

$$\therefore \qquad \delta = \frac{\sigma L}{E}$$

Also $\text{force}=\text{stress}\times\text{area}=\sigma A$

Where A is the cross sectional area

Therefore Rule 1 becomes

$$\frac{\sigma_S L}{E_S} + \frac{\sigma_B L}{E_B} = (\alpha_B = \alpha_S)Lt$$

Rule 2 becomes

$$\sigma_S A_S = \sigma_B A_B$$

As a result, we have two equations with unknown parameters σ_s and σ_b, and we can determine how much these stresses are (Thanos & Rundel, 1995).

4.7. BENDING MOMENT AND SHEARING FORCE DIAGRAMS

The shearing force can be described as the sum of the forces exerted on either side of any region on a beam bearing transverse loads. The bending moment at any region is measured by adding the moments of the loads exerted on that section, again taken from each side. A practical sign convention must be chosen so that the bending-moment values and shearing force calculated on either side of the region have the same sign and magnitude. The images below depict this concept (Zhang et al., 2014).

Bending-moment diagrams and shearing force illustrate how these values change across a beam's length under any fixed stress situation.

There are likely to be resultant forces along any side of every segment in a beam possessing transverse loads. For equilibrium, these forces ought to be opposite and equal, and their combined effect attempts to shear the segment in two different ways depicted in Figure below (Song, 2013).

The algebraic mean of the forces applied to one side of the segment is consequently used to define the shearing force (S.F.) at the segment. Selecting a side is a question of convenience, but a convenient sign convention should be used for the value received on both sides to share the same sign and magnitude (Rahal, 2000).

4.7.1. Sign Convention for Shearing Force (S.F.)

Positive forces are upward to the left or downward to the right of a segment. As a result, the S.F. system at X-X is shown positively in below image (a) and negatively in Figure 3.1(b).

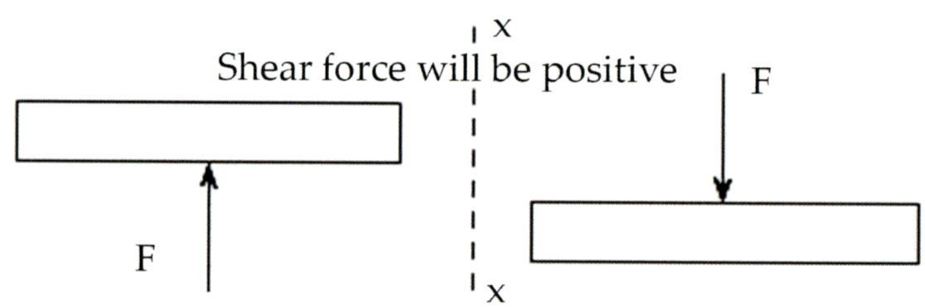

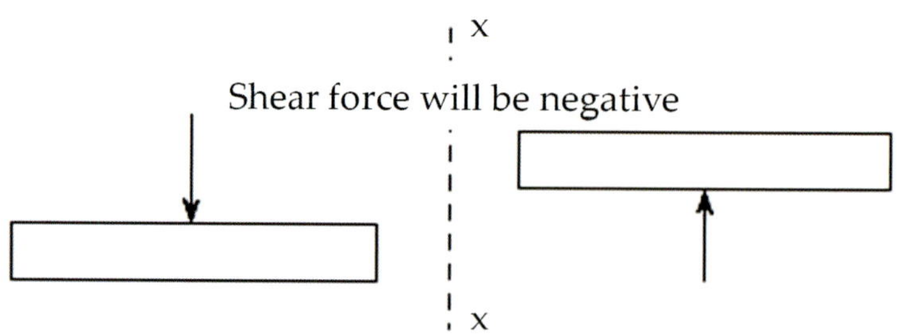

Every segment of the beam will also experience bending, or a resultant B.M., the sum of the moments of all of the separate loads, in addition to shear. Also, the values along either side of the segment must be identical to be in equilibrium.

This leads to the definition of the bending moment (B.M.) as the total of the moments of the forces about the segment, measured from any side of the segment. A practical sign standard must be selected for S.F (Brooks, 1928).

4.7.2. Sign Convention for Bending Moment (B.M.)

Moments are positive at clockwise to the left and anticlockwise to the right. Thus, Figure 4.15(a) depicts a positive B.M. system that causes the beam at X-X to sag, and Figure 4.15(b) depicts a negative B.M. system with its corresponding snagging beam.

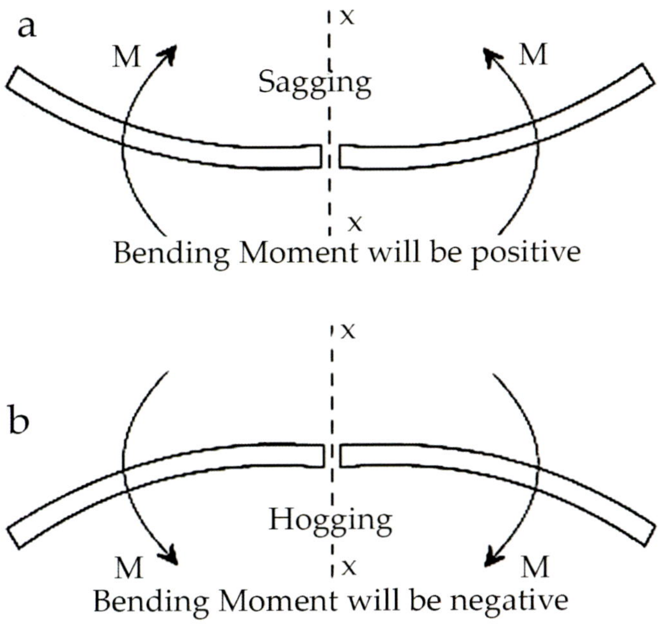

Figure 4.15. Image of sign conventions of bending moment.

Source: Harikesh Kumar, Creative Commons License.

Although the previously mentioned sign conventions for B.M and S.F. are somewhat imprecise and might be reversed, it must be mentioned that the systems selected here are the only ones that produce the mathematically accurate signs for beam and slope deflections in later studies and are, therefore strongly advised (Stephens et al., 1972).

As a result, they are graphs of B.M and S.F. values written on the beam as a basis, and they show, in the early stages of design, the points on the beam that are subjected to the largest shear or bending stresses and, as a result, which might need further strengthening or attention.

It is crucial to note that structures may encounter two main types of loading: concentrated loads and distributed loads. Because all practical loading systems should be administered over a finite area, the former is expected to work at a certain point and simultaneously create an oversimplification. Yet, this area is supposed to be minimal because the load can legitimately be considered to work at a single location. Uniformly distributed loads are presumed to work over all or a portion of the beam and are typically equally distributed (Hulse & Cain, 2000).

On rare occasions, the dispersion is not constant but may fluctuate linearly along the loaded part or exhibit a more complicated dispersion structure (Figure 4.16).

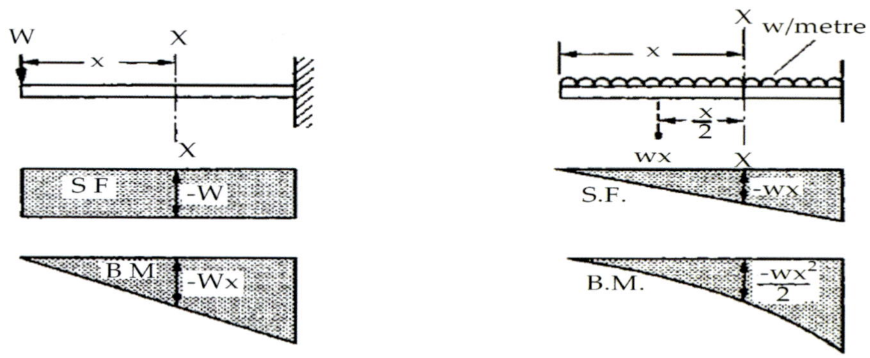

Figure 4.16. Diagram of S.F-B.M diagram for standard cases.

Source: E. J. Hearn, Creative Commons License.

Hence, the S.F. at any segment X-X, located x from the free end, is S.F. = – W in an example of a beam bearing a concentrated load W at the end (Figure 4.16). This holds regardless of the x-value, leading to the S.F. diagram's transformation into a rectangle. The B.M. is – Wx at the same segment X-X and will rise linearly with x. Thus, the B.M. diagram is a triangle. The S.F. at X-X is the net load to one side of X-X, or -wx if the beam currently carries an evenly distributed load. Hence, the S.F. diagram assumes a triangle shape and grows to a maximum value of – WL at the base. The B.M. at X-X is produced by considering the load to the left of X-X as a concentrated load with a comparable value applied at the center of gravity (Cain et al., 1990).

$$\text{B.M. at } X - X = -wx\frac{x}{x} = \frac{-wx^2}{2}$$

4.8. BENDING THEORY

If we now consider a beam initially unstressed and subjected to a constant B.M.along its length, i.e. pure bending, as would be obtained by applying equal couples at each end, it will bend to a radius R as shown in Figure(b) below. The top fibers of the beam will experience tension as a consequence of this bending, whereas the bottom fibers will experience compression. So, it seems sensible to assume that there are points with zero stress between the two. The neutral axis is the location of all of these points. Then, we measure the radius of R curvature to this axis (Costello & Butson, 1982). The N.A. is the symmetrical axis for symmetrical segments, although regardless of the segment, the N.A. will always run through the centroid.

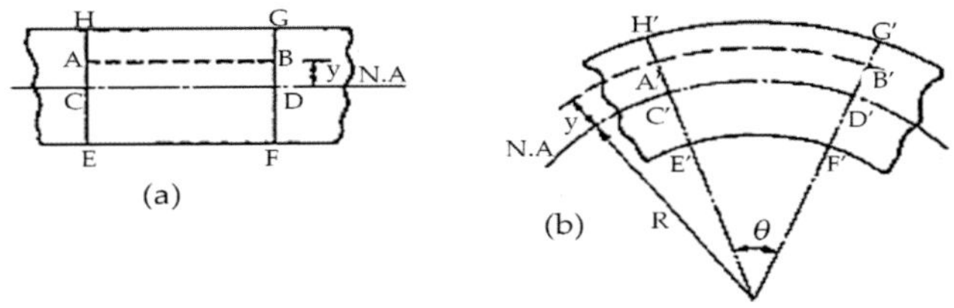

Assume two initially parallel cross-sections of a beam, GF and HE (Figure 4.3). These portions are believed to stay plane when the beam is bent (Figure 4.2(b)), meaning that HE and GF,' the sections' final locations, are still straight lines. Then they will displace an angle θ.

Consider a few AB fibers in a material x distance from the N.A. When the beam is curved, it will expand to A'B' (Yu & Zhang, 1996).

But AB=CD, and, since the N.A is unstressed, CD=C'D'

Equation the two equations for strain,

$$\frac{\sigma}{E} = \frac{y}{R}$$

$$\frac{\sigma}{y} = \frac{E}{R} \qquad (4.11)$$

Consider now a cross-section of the beam. From Eqn. (4.11) the stress on fiber at distance y from the NA, is:

$$\sigma = \frac{E}{R} y \qquad (4.12)$$

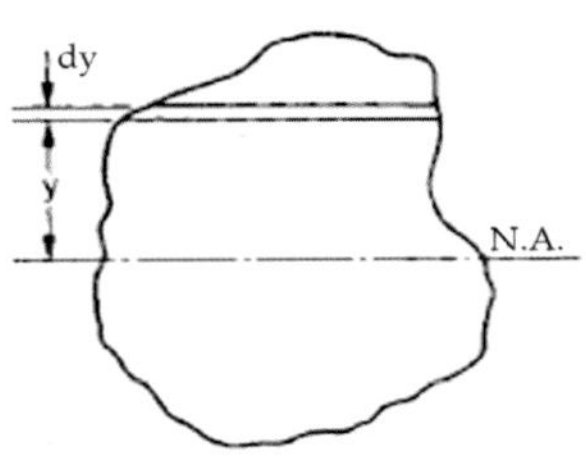

CHAPTER 4

The force on strip of area 6A is:

$$F = \sigma\delta A = \frac{E}{R}y\delta A$$

The moment about the N.A. is:

$$Fy = \frac{E}{R}y^2\delta A$$

Thus, the total moment for the complete cross-section is:

$$M = \sum \frac{E}{R}y^2\delta A$$

$$= \frac{E}{R}\Sigma y^2\delta A$$

Here E and R are constant.

The term $\Sigma y^2\delta A$ is known as the second moment of the cross-sectional area and given the symbol I (Petrangeli et al., 1999).

$$M = \frac{E}{R}I \text{ and } \frac{M}{I} = \frac{E}{R} \quad (4.13)$$

By connecting Eqns. (4.11) and (4.12), we will have an equation for the bending theory:

$$\frac{M}{I} = \frac{\sigma}{y} = \frac{E}{R} \quad (4.14)$$

From Eqn. (4.13), it is evident that the I, M, and E values remain constant, which means that the radius of curvature of the curved beam will likewise stay constant if the beam is of the constant segment, the material is homogeneous, and the induced moment is constant. The term "circular bending" is frequently used because beams will curve into circular arcs for pure bending of homogeneous segments. The radius of curvature at which a given moment M curves a beam is determined by Eqn. (4.13) as follows (Wang et al., 2008):

$$R = \frac{EI}{M}$$

The parameter EI is frequently referred to as the flexural rigidity or flexural beam stiffness because the radius of curvature is a direct indicator of the degree of flexibility of the beam (the larger the R-value, the smaller the bending and the larger the rigidity). The EI values thus make it simple to compare the relative bending stiffness of beam sections.

It should be noted that the evidence presented above relied on the presumption that there was only pure bending and no shear. It is evident from the study of the previous chapter that shear and bending typically occur concurrently at most places in practical beam-loading instances (Jolicoeur & Cardou, 1996).

But a closer look at the B.M. and S.F. diagrams reveals that the S.F. is, in practice, always zero when the B.M. is at its maximum. In all but a small fraction of unique loading circumstances, it will eventually be demonstrated that bending generates the most stress, ensuring that beams constructed using the simple bending theory and the optimum B.M. are typically stronger at other points than necessary (Frostig & Baruch, 1996).

4.9. NEUTRAL AXIS

As was already mentioned, it is obvious that there should be a point within the cross-section of a beam where the stress varies sign, like, with zero stress, and this is known as the neutral axis; if, during curving, one beam surface is exposed to tension and the reverse surface to compression.

Further, Eqn. (4.14) may be re-written as (Eltaher et al., 2014):

$$\sigma = (M/I)y \qquad (4.15)$$

Substituting from Eqn. (4.11)

As y is always calculated from the NA, this integral equal the initial moment of the area of the beam's cross-section about the NA. Since this is the fundamental requirement of the centroid, the only first moment of area for the cross-section that is zero at the moment is about an axis through the section's centroid. The neutral axis must consequently always run across the centroid It must be considered that other criteria should be fulfilled when stresses cross into the plastic region of the materials in concern; this criterion only applicable while stresses are held within the elastic zone of the material in concern (Shi et al., 2014).

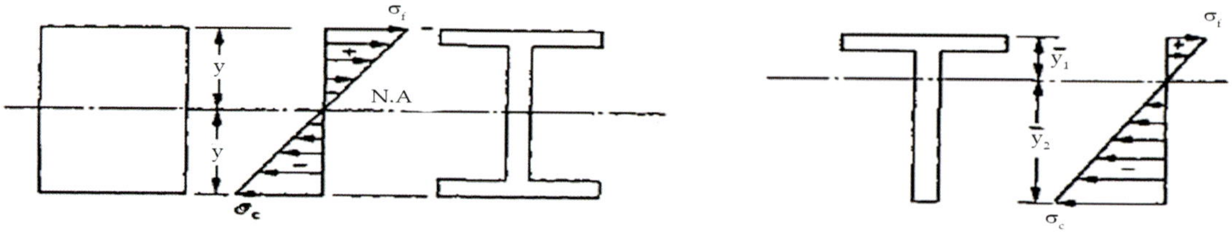

Figure 4.4 depicts common stress distributions in bending. It is clear that relative to regions farthest from the axis, the material close to the N.A. is consistently exposed to relatively modest stresses. Using segments with large areas far from the N.A. is advised to get the greatest bending resistance. Because of this, beams with I- or T-segments are highly preferred in modern engineering applications like beams because bending plays a significant role. These beams have significant moments of an area about a single axis, and as long as this axis is used for bending, they will be very resistant to bending forces (Soman et al., 2018).

4.10. BUILT-IN BEAMS

The terms "built-in," "encastred," and "encastri" are used to describe a beam when both ends are firmly connected. Such beams are often handled using Macaulay's method or a revised version of Mohr's area-moment method. It is considered that built-in beams have a zero slope at each end, resulting in a minimal change in slope over the span (Okafor et al., 1996). Thus, using Mohr's first theorem

area of M/EI diagram across the span=0

or constant for uniform beam, is Ei, also

Area of B.M. figure = 0

The deviation of one end concerning the other is zero if both ends are equal.

Thus, according to Mohr's second theorem:

First moment of area of M/EI diagram about one end=0

and, if EI is constant,

First moment of area of B.M. diagram about one end=0 (4.16)

It is advantageous to divide the B.M. figure for the designed beam into two pieces to use these equations:

- The diagram representing the end moments that have to be provided at the ends to maintain the slope zero is called the fixing-moment figure.

- The diagram representing the loading, supposing supported ends, is called the free-moment diagram (Wittwer & Howell, 2004).

4.11. PROS AND CONS OF BUILT-IN BEAMS

Designed beams hold smaller maximum B.M.s. They have relatively small deformations than the resultant supported beams with the same applied loads, meaning that designed beams are stronger and stiffer—providing that perfect end fixing can be accomplished. While this might suggest that designed beams must be employed wherever possible, this is not the situation in real-world applications. The following are the main causes (Fewins, 2004):

1. Costs go up because the precise alignment of the supports and fastening of the ends are required for erection.
2. Large stresses can be created by even slight sinking of each support.
3. Temperature fluctuations can also create large strains.
4. When rolling loads are introduced into an application, the end fixings are typically sensitive to B.M. variations and vibrations like bridges.

Moreover, if hinge joints are employed at positions on the beam at zero B.M., at contra flexure or inflection points, these drawbacks can be lessened. The structure is frequently called the double-cantilever architecture since the beam is functionally carried on two end cantilevers as a center beam. The beam is then adapted to temperature fluctuations (Figure 6.5) (Zhdanov et al., 2016).

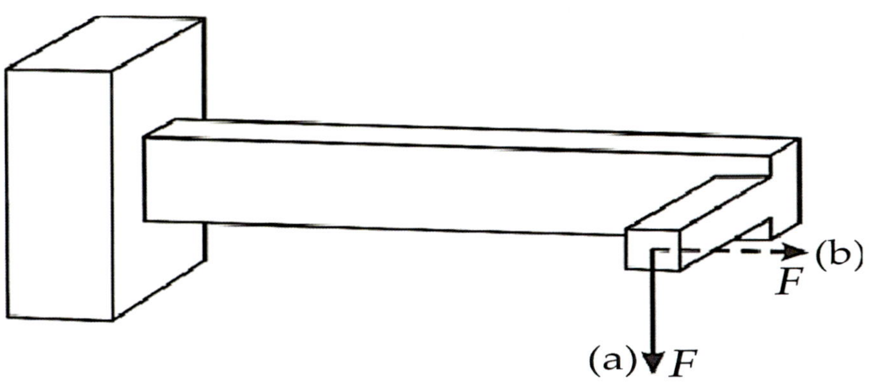

Think about a beam AB initially offloaded with both ends at the equal height. The moments must be as illustrated in Figure, if the slope stays horizontal for each end when B travels through a length of 6 concerning end A.

$R_A \times L = M_A + M_B$

And by symmetry

$M_A = M_B = M$

R_A =

R_B =

The bending of A concerning B is equivalent to the first moment of the region of the B.M. diagram now that Mohr's second theorem has been established $A \times 1/EI$. in the directions depicted in Figure 4.17.

δ= [(-M X) + (M X)] (4.17)

(-1+5)= (4.18)

M= and R_A= R_B= (4.19)

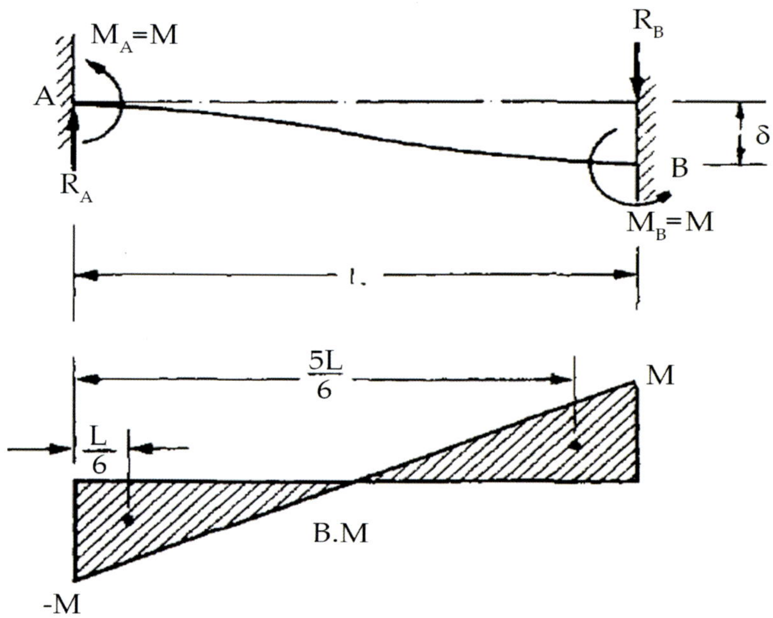

Figure 4.17. Illustration of effect of support movement on B.M.s.

Source: University of Warwick, Creative Commons License.

Also, these numbers will show how the fixing moments and end responses of a beam under load variation as one end sinks concerning the other (Mitra et al., 2005).

4.12. STRUCTURAL STABILITY

The capability of a structure or system to keep its shape and functionality under different internal or external forces, like vibrations, loads, temperature fluctuations, or material deterioration, is known as structural stability. A system that can withstand collapse or failure while maintaining stability and functionality over time is considered structurally stable (Alfutov & Wauer, 2002).

Structural stability is essential in many disciplines, including physics, biology, and engineering architecture, where complex structures or systems must be created, assessed, and maximized to suit certain criteria and restrictions. A fundamentally sound system can guarantee safety, durability, dependability, and efficiency while lowering the likelihood of unanticipated failures (Simitses & Hodges, 2006).

Below are some basic aspects that provide structural stability:

1. **Strength:** A structure's strength is determined by its capacity to bear outside pressure without deforming or breaking. When faced with loads like wind, weight, vibrations, and collision, a rigid structure should be capable of withstanding them without cracking or collapsing. Geometry, material characteristics, and structural layout affect a structure's strength (Du et al., 2005).
2. **Stiffness:** A structure's rigidity is defined as its tolerance to distortion under load. Under load, a rigid construction will keep its size and shape and won't deform too much. Especially in dynamic or vibratory situations, a structure must be stiff to retain its stability (Hoshiya & Saito, 1984).
3. **Damping:** The term "damping" describes a structure's capacity to release energy and lessen the intensity of vibrations. In constructions subjected to repeated loads or vibration, damping is necessary to avoid fatigue and resonance breakdown (Bergman et al., 1997).
4. **Buckling:** When a thin structure, like a beam or column, breaks as a result of too much compressive stress, it is said to have buckled. The structure's stiffness, strength, or geometrical layout could be improved to minimize buckling in addition to the applied loads (Housner et al., 1997).
5. **Fatigue:** When a structure is subjected to cyclic stress or recurrent loading, it is considered fatigued. The structure may develop cracks, breaches, or collapse due to fatigue breakdown. Fatigue can be reduced by employing materials with high fatigue tolerance and designing the structure with sufficient damping, strength, stiffness, and geometrical features (Bucher & Lin, 1988).
6. **Durability:** The tendency of a building to sustain its functionality and coherence over time despite diverse climatic or operating situations is referred to as durability. Weathering, material aging, corrosion, and maintenance procedures are some variables that can affect durability (Ulm et al., 2000).

An area of mechanics called as structural stability measures the structures' response to pressure. A material has the propensity to lose stiffness, undergo a discernible variation in geometry, and become unsteady when exposed to a large enough compressive force. When instability happens, the structure cannot stay in an equilibrium state and cannot support the applied loads (Kollár, 1999).

4.13. INSTABILITY AND ITS TYPES

Limit point instability, finite disturbance instability, Bifurcation instability, and snap-through instability are common categories of instability. The following is a summary of each.

4.13.1. Bifurcation Instability

Bifurcation instability is when distortion occurring in one direction abruptly shifts to another. An illustration would be a simple symmetrical column subjected to a concentric compressive load. The column initially suffers axial distortion or shortening in the direction in which the force is exerted. At some point, when the applied force progressively rises, the deformation mode abruptly changes from axial to lateral, causing the column to bow perpendicular to the direction of the applied force (Chambon et al., 2004).

4.13.2. Limit Point Instability

This situation is called limit point instability when a single distortion mode persists throughout the load history. The deformation grows as the load increases from the beginning of loading to the eventual breakdown. The limit load is the greatest weight that a structure can support before failing. Geometrically flawed columns subjected to circumferential compressive forces and frames subject to eccentric gravity loadings to the longitudinal planes of the columns are samples of structures that display limit point instability (Gillman et al., 2018).

4.13.3. Finite Disturbance Instability

When a compressive force is exerted along a thin-walled cylinder shell's longitudinal or axial direction, finite disturbance instability happens. The load-deflection curve increases to the critical load Ncr and then abruptly falls to a lesser value in order for the structure to achieve stability. Donnell and Wan have demonstrated that the value of Ncr is extremely dependent on the initial geometrical flaws in the shell. Even the smallest flaws significantly lower Ncr (McIntire & Shepherd, 1987).

4.13.4. Snap-Through Instability

A sharp rise in distortion defines snap-through instability while the load remains constant. When a load exceeds a certain level, some processes, including in-plane crumpling of a simple truss or arch subjected to a transverse load and collapsing of a shallow spherical cap subjected to a radial load, can only sustain equilibrium if the displacement snaps abruptly from point A to point B, as indicated by the solid horizontal line.

The system must be exposed to an ideal displacement-regulated condition in order to see the dotted curving line, which reflects an unstable equilibrium state (Liu et al., 2017).

ACTIVITY 1.1.

A bridge is being constructed over a river. The bridge's engineer is concerned about the weight that the bridge can handle without breaking. How can the principles of mechanics of materials be applied to determine the maximum weight the bridge can safely support?

4.14. SUMMARY

-The chapter on mechanics of materials focuses on the study of the behavior of solid materials under different loading conditions. Stress and strain are the fundamental concepts that help in determining the mechanical properties of materials. The tensile test is a common method for determining the mechanical properties of materials. Materials can be classified as brittle or ductile based on their ability to withstand deformation without fracturing. Poisson's ratio is a material property that relates the lateral and longitudinal strains of a material. Shear stress and strain are important in determining the behavior of materials under shear loading conditions. Compound bars are structures made of multiple materials that are connected in series or parallel, and their behavior can be analyzed using stress and strain calculations. External loads and temperature fluctuations can cause stress and strain in compound bars. Bending moment and shearing force diagrams are graphical representations of the forces acting on a beam under bending and shear loading conditions. Bending theory is used to determine stresses and strains in beams, and the neutral axis is a location in a beam where the stress is zero. Built-in beams are structures that are rigidly connected to their supports, and they have their own unique advantages and disadvantages. Instability is an important concept in mechanics of materials, and it can take different forms such as buckling, torsional instability, and shear instability.

REVIEW QUESTIONS

1. What are the fundamental concepts used in mechanics of materials?
2. Describe the tensile test and its importance in determining mechanical properties of materials.
3. Differentiate between brittle and ductile materials.
4. What is Poisson's ratio and how is it calculated?
5. What are the different types of stresses and strains that can be experienced by a material?
6. How are compound bars analyzed for stress and strain?
7. What is the effect of external loads and temperature fluctuations on compound bars?
8. How are bending moment and shearing force diagrams used to analyze beams under loading conditions?
9. What is the neutral axis of a beam, and how is it determined?
10. What are built-in beams and what are their advantages and disadvantages?
11. What is instability in mechanics of materials, and what are the different types of instability?

MULTIPLE CHOICE QUESTIONS

1. What is the fundamental concept used to determine the mechanical properties of materials?

a. Temperature
b. Pressure
c. Stress and strain
d. Density

2. What type of test is commonly used to determine the mechanical properties of materials?

a. Shear test
b. Compression test
c. Tensile test
d. Bending test

3. What is the main difference between brittle and ductile materials?

a. Ductile materials have higher density than brittle materials.
b. Brittle materials have higher strength than ductile materials.
c. Ductile materials can undergo plastic deformation while brittle materials cannot.
d. Brittle materials can absorb more energy than ductile materials.

4. What is Poisson's ratio?

a. A measure of the material's ability to withstand deformation without fracturing.
b. A measure of the material's strength under compression.
c. A material property that relates the lateral and longitudinal strains of a material.
d. A measure of the material's ability to resist shear stress.

5. What is shear stress?

a. The force that opposes the deformation of a material.
b. The force that causes a material to compress.
c. The force that causes two adjacent planes in a material to slide past each other.
d. The force that causes a material to bend.

6. What is a compound bar?

a. A structure made of multiple materials that are connected in series or parallel.
b. A type of bar that is designed to withstand torsional loading.
c. A bar that is designed to bend under load.
d. A type of bar that is used to support heavy loads.

7. What is the neutral axis of a beam?

a. The location in a beam where the stress is zero.

b. The location in a beam where the shear stress is maximum.

c. The location in a beam where the bending stress is maximum.

d. The location in a beam where the deflection is maximum.

8. What is instability in mechanics of materials?

a. The ability of a material to withstand deformation without fracturing.

b. The ability of a material to resist shear stress.

c. The tendency of a structure to collapse under load.

d. The tendency of a material to undergo plastic deformation.

Answers to Multiple Choice Questions

1. (c); 2. (c); 3. (c); 4. (c); 5. (c); 6. (a); 7. (a); 8. (c)

REFERENCES

1. Alfutov, N. A., & Wauer, J., (2002). Stability of elastic structures. Foundations of engineering mechanics. Appl. Mech. Rev., 55(2), B32.
2. Aquaro, D., & Fontani, E., (2001). Erosion of ductile and brittle materials. Meccanica, 36, 651–661.
3. Bažant, Z. P., (2000). Structural stability. International Journal of Solids and Structures, 37(1, 2), 55–67.
4. Bergman, L. A., Shinozuka, M., Bucher, C. G., Sobczyk, K., Dasgupta, G., Spanos, P. D., & Zhang, R., (1997). A state-of-the-art report on computational stochastic mechanics. Probabilistic Engineering Mechanics, 12(4), 197–321.
5. Bert, C. W., & Zeng, H., (2001). Analysis of axial vibration of compound bars by several different methods. In: International Design Engineering Technical Conferences and Computers and Information in Engineering Conference American Society of Mechanical Engineers (Vol. 80296, pp. 2215–2219).
6. Brooks, W. H., (1928). LXXVII. Sign conventions applied to flexing problems. The London, Edinburgh, and Dublin Philosophical Magazine and Journal of Science, 5(30), 749–756.
7. Bucher, C. G., & Lin, Y. K., (1988). Stochastic stability of bridges considering coupled modes. Journal of Engineering Mechanics, 114(12), 2055–2071.
8. Cabezas, E. E., & Celentano, D. J., (2004). Experimental and numerical analysis of the tensile test using sheet specimens. Finite Elements in Analysis and Design, 40(5, 6), 555–575.

9. Cain, J., Hulse, R., Cain, J., & Hulse, R., (1990). Shearing forces and bending moments. Structural Mechanics, 2, 99–136.
10. Chambon, R., Caillerie, D., & Viggiani, G., (2004). Loss of uniqueness and bifurcation vs instability: Some remarks. Revue Française de Génie Civil, 8(5, 6), 517–535.
11. Chen, T., & Weng, I. S., (2001). Torsion of a circular compound bar with imperfect interface. J. Appl. Mech., 68(6), 955–958.
12. Costello, G. A., & Butson, G. J., (1982). Simplified bending theory for wire rope. Journal of the Engineering Mechanics Division, 108(2), 219–227.
13. Daniel, I. M., Ishai, O., Daniel, I. M., & Daniel, I., (2006). Engineering Mechanics of Composite Materials (Vol. 1, pp. 1–30). New York: Oxford university press.
14. Dell'Isola, F., Sciarra, G., & Vidoli, S., (2009). Generalized Hooke's law for isotropic second gradient materials. Proceedings of the Royal Society A: Mathematical, Physical and Engineering Sciences, 465(2107), 2177–2196.
15. Du, S., Ellingwood, B. R., & Cox, J. V., (2005). Solution methods and initialization techniques in SFE analysis of structural stability. Probabilistic Engineering Mechanics, 20(2), 179–187.
16. Duc, N. H., Anh, D. K., & Brommer, P. E., (2002). Metamagnetism, giant magnetoresistance and magnetocaloric effects in RCo2-based compounds in the vicinity of the curie temperature. Physica B: Condensed Matter, 319(1–4), 1–8.
17. Eltaher, M. A., Abdelrahman, A. A., Al-Nabawy, A., Khater, M., & Mansour, A., (2014). Vibration of nonlinear graduation of nano-Timoshenko beam considering the neutral axis position. Applied Mathematics and Computation, 235, 512–529.
18. Emri, I., Voloshin, A., Emri, I., & Voloshin, A., (2016). Compound structures. Statics: Learning from Engineering Examples, 3(1), 457–488.
19. Fang, S., Li, L., Luo, Z., Fang, Z., Huang, D., Liu, F., & Xiong, Z., (2023). Novel FRP interlocking multi-spiral reinforced-seawater sea-sand concrete square columns with longitudinal hybrid FRP–steel bars: Monotonic and cyclic axial compressive behaviors. Composite Structures, 305, 116487.
20. Fewins, C., (2004). The pros and cons of different construction systems. Home Building & Renovating, 1, 1–11.
21. Fischer, I., Pichler, B., Lach, E., Terner, C., Barraud, E., & Britz, F., (2014). Compressive strength of cement paste as a function of loading rate: Experiments and engineering mechanics analysis. Cement and Concrete Research, 58, 186–200.
22. Frostig, Y., & Baruch, M., (1996). Localized load effects in high-order bending of sandwich panels with flexible core. Journal of Engineering Mechanics, 122(11), 1069–1076.
23. Gianola, D. S., & Eberl, C., (2009). Micro-and nanoscale tensile testing of materials. JOM, 61, 24–35.
24. Gillman, A., Fuchi, K., & Buskohl, P. R., (2018). Truss-based nonlinear mechanical analysis for origami structures exhibiting bifurcation and limit point instabilities. International Journal of Solids and Structures, 147, 80–93.

25. Gross, D., Hauger, W., Schröder, J., Wall, W. A., & Bonet, J., (2011). Engineering Mechanics (Vol. 2, pp. 1–23). Springer.
26. Hannon, A., & Tiernan, P., (2008). A review of planar biaxial tensile test systems for sheet metal. Journal of Materials Processing Technology, 198(1–3), 1–13.
27. Hart, E. W., (1967). Theory of the tensile test. Acta Metallurgica, 15(2), 351–355.
28. Hartsuijker, C., & Welleman, J. W., (2007). Engineering Mechanics: Stresses, Strains, Displacements (Vol. 2, pp. 1–50). Springer Science & Business Media.
29. Hoshiya, M., & Saito, E., (1984). Structural identification by extended Kalman filter. Journal of Engineering Mechanics, 110(12), 1757–1770.
30. Housner, G., Bergman, L. A., Caughey, T. K., Chassiakos, A. G., Claus, R. O., Masri, S. F., & Yao, J. T., (1997). Structural control: Past, present, and future. Journal of Engineering Mechanics, 123(9), 897–971.
31. Hulse, R., & Cain, J., (2000). Shearing forces and bending moments. In: Structural Mechanics (Vol. 1, pp. 99–136). Red Globe Press, London.
32. Jiao, G., & Moan, T., (1990). Probabilistic analysis of fatigue due to Gaussian load processes. Probabilistic Engineering Mechanics, 5(2), 76–83.
33. Jolicoeur, C., & Cardou, A., (1996). Semicontinuous mathematical model for bending of multilayered wire strands. Journal of Engineering Mechanics, 122(7), 643–650.
34. Krishnan, S. R., Su, C. J., Xie, Z., Patel, M., Madhvapathy, S. R., Xu, Y., & Rogers, J. A., (2018). Wireless, battery-free epidermal electronics for continuous, quantitative, multimodal thermal characterization of skin. Small, 14(47), 1803192.
35. Kubıcek, M., & Florian, Z., (2009). Stress strain analysis of knee joint. Engineering Mechanics, 16(5), 315–322.
36. LaVan, D. A., & Sharpe, W. N., (1999). Tensile testing of microsamples. Experimental Mechanics, 39, 210–216.
37. Lee, T., & Lakes, R. S., (1997). Anisotropic polyurethane foam with Poisson's ratio greater than 1. Journal of Materials Science, 32, 2397–2401.
38. Li, J., Cui, X. Z., Qi, H., Jin, Q., Su, J. W., Wang, Y. L., & Zhang, X. N., (2021). Laboratory tests on engineering properties of a new negative-Poisson′ s-ratio geobelt. Construction and Building Materials, 297, 1–75.
39. Liu, N., Plucinsky, P., & Jeffers, A. E., (2017). Combining load-controlled and displacement-controlled algorithms to model thermal-mechanical snap-through instabilities in structures. Journal of Engineering Mechanics, 143(8), 11–63.
40. Lord, J. D., & Morrell, R. M., (2010). Elastic modulus measurement—Obtaining reliable data from the tensile test. Metrologia, 47(2), S41–S45.
41. Masoero, E., Wittel, F. K., Herrmann, H. J., & Chiaia, B. M., (2010). Progressive collapse mechanisms of brittle and ductile framed structures. Journal of Engineering Mechanics, 136(8), 987–995.
42. McGarry, J. G., Klein-Nulend, J., Mullender, M. G., & Prendergast, P. J., (2005). A comparison of strain and fluid shear stress in stimulating bone cell responses—a computational and

experimental study. The FASEB Journal, 19(3), 1–22.
43. McIntyre, M. E., & Shepherd, T. G., (1987). An exact local conservation theorem for finite-amplitude disturbances to non-parallel shear flows, with remarks on Hamiltonian structure and on Arnold's stability theorems. Journal of Fluid Mechanics, 181, 527–565.
44. Mitra, S., Seifert, N., Zhang, M., Shi, Q., & Kim, K. S., (2005). Robust system design with built-in soft-error resilience. Computer, 38(2), 43–52.
45. Muhanna, R. L., Zhang, H., & Mullen, R. L., (2007). Interval finite elements as a basis for generalized models of uncertainty in engineering mechanics. Reliable Computing, 13, 173–194.
46. Neggers, J., Hoefnagels, J. P. M., Hild, F., Roux, S., & Geers, M. G. D., (2014). Direct stress-strain measurements from bulged membranes using topography image correlation. Experimental Mechanics, 54, 717–727.
47. Nordmann, J., Aßmus, M., Glüge, R., & Altenbach, H., (2020). On the derivation of Hooke's law for plane state conditions. Technische Mechanik-European Journal of Engineering Mechanics, 40(2), 160–174.
48. Okafor, A. C., Chandrashekhara, K., & Jiang, Y. P., (1996). Delamination prediction in composite beams with built-in piezoelectric devices using modal analysis and neural network. Smart Materials and Structures, 5(3), 338.
49. Petrangeli, M., Pinto, P. E., & Ciampi, V., (1999). Fiber element for cyclic bending and shear of RC structures. I: Theory. Journal of Engineering Mechanics, 125(9), 994–1001.
50. Phillips, C. E., & Heywood, R. B., (1951). The size effect in fatigue of plain and notched steel specimens loaded under reversed direct stress. Proceedings of the Institution of Mechanical Engineers, 165(1), 113–124.
51. Rahal, K. N., (2000). Shear strength of reinforced concrete: Part II—Beams subjected to shear, bending moment, and axial load. Structural Journal, 97(2), 219–224.
52. Reinhardt, H. W., Cornelissen, H. A., & Hordijk, D. A., (1986). Tensile tests and failure analysis of concrete. Journal of Structural Engineering, 112(11), 2462–2477.
53. Rychlewski, J., (1984). On Hooke's law. Journal of Applied Mathematics and Mechanics, 48(3), 303–314.
54. Saadatmanesh, H., Ehsani, M. R., & Li, M. W., (1994). Strength and ductility of concrete columns externally reinforced with fiber composite straps. Structural Journal, 91(4), 434–447.
55. Sabik, A., (2018). Direct shear stress vs strain relation for fiber reinforced composites. Composites Part B: Engineering, 139, 24–30.
56. Sales, B. C., & Wohlleben, D. K., (1975). Susceptibility of interconfiguration-fluctuation compounds. Physical Review Letters, 35(18), 1240.
57. Senitsky, Y. E., & Ishutin, A. S., (2015). Overall stability of compressed compound bars in variable cross section. Journal of Samara State Technical University, Ser. Physical and Mathematical Sciences, 19(2), 341–357.

58. Shi, Y., Rogers, J. A., Gao, C., & Huang, Y., (2014). Multiple neutral axes in bending of a multiple-layer beam with extremely different elastic properties. Journal of Applied Mechanics, 81(11), 111–222.

59. Simitses, G., & Hodges, D. H., (2006). Fundamentals of Structural Stability (Vol. 1, pp. 1–17). Butterworth-Heinemann.

60. Soman, R., Malinowski, P., Majewska, K., Mieloszyk, M., & Ostachowicz, W., (2018). Kalman filter based neutral axis tracking in composites under varying temperature conditions. Mechanical Systems and Signal Processing, 110, 485–498.

61. Song, X. J., (2013). Quick drawing of internal forces diagram of beam. In: Applied Mechanics and Materials (Vol. 275, pp. 128–131). Trans Tech Publications Ltd.

62. Stephens, R. C., Ward, J. J., Stephens, R. C., & Ward, J. J., (1972). Shearing force and bending moment diagrams. Applied Mechanics, 1(1),108–118.

63. Taylor, G. I., & Elam, C. F., (1923). Bakerian lecture: The distortion of an aluminum crystal during a tensile test. Proceedings of the Royal Society of London. Series A, Containing Papers of a Mathematical and Physical Character, 102(719), 643–667.

64. Thanos, C. A., & Rundel, P. W., (1995). Fire-followers in chaparral: Nitrogenous compounds trigger seed germination. Journal of Ecology, 1(2), 207–216.

65. Tvergaard, V., Needleman, A., & Lo, K. K., (1981). Flow localization in the plane strain tensile test. Journal of the Mechanics and Physics of Solids, 29(2), 115–142.

66. Ulm, F. J., Coussy, O., Kefei, L., & Larive, C., (2000). Thermo-chemo-mechanics of ASR expansion in concrete structures. Journal of Engineering Mechanics, 126(3), 233–242.

67. Wang, C. M., Kitipornchai, S., Lim, C. W., & Eisenberger, M., (2008). Beam bending solutions based on nonlocal Timoshenko beam theory. Journal of Engineering Mechanics, 134(6), 475–481.

68. Wingender, D., & Balzani, D., (2022). Simulation of crack propagation based on eigenerosion in brittle and ductile materials subject to finite strains. Archive of Applied Mechanics, 92(4), 1199–1221.

69. Wittwer, J. W., & Howell, L. L., (2004). Mitigating the effects of local flexibility at the built-in ends of cantilever beams. J. Appl. Mech., 71(5), 748–751.

70. Yang, G., Kabel, J., Van, R. B., Odgaard, A., Huiskes, R. I. K., & Cown, S. C., (1998). The anisotropic Hooke's law for cancellous bone and wood. Journal of Elasticity, 53, 125–146.

71. Yu, T., & Zhang, L., (1996). Plastic Bending: Theory and Applications (Vol. 2, pp. 1–29). World Scientific.

72. Zhang, H. W., Shen, Y., & Zhang, S. W., (2014). To solve the load of beam by shearing force diagram and bending moment diagram-reverse thinking in strength of materials. In: Applied Mechanics and Materials (Vol. 488, pp. 546–549). Trans Tech Publications Ltd.

73. Zhdanov, B. V., Rotondaro, M. D., Shaffer, M. K., & Knize, R. J., (2016). Low-pressure cesium and potassium diode pumped alkali lasers: Pros and cons. Optical Engineering, 55(2), 026105–026105.
74. Zhu, T., & Huang, D., (2019). Experimental investigation of the shear mechanical behavior of sandstone under unloading normal stress. International Journal of Rock Mechanics and Mining Sciences, 114, 186–194.

CHAPTER 5

VIRTUAL WORK AND ENERGY

UNIT INTRODUCTION

Virtual work and energy are significant ideas in engineering mechanics essential to comprehend how structures behave under strain. The deflection, stability, and other characteristics of structures, like plates, frames, and beams, are examined using these principles.

According to the principles of virtual work, the intrinsic work a structure does on its own due to distortion is equivalent to the exterior work that applied load exerts on a structure. This method can ascertain a system's equilibrium by offering a clear and elegant solution to tricky structural issues.

The virtual work concept and the energy approach are effective methods for calculating the sum of the potential energy of the structure and its reaction to the load. This chapter gives a basis for comprehending the response of mechanical systems by introducing the ideas of virtual labor and energy and their usage in structural analysis (Georgiev et al., 2014) (Figure 5.1).

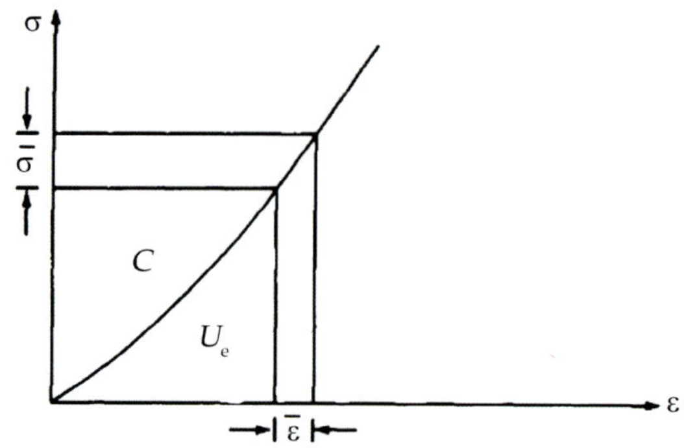

Figure 5.1. Graphical representation of virtual work.

Source: Science Direct, Creative Commons License.

Learning Objectives

At the end of this chapter, readers will be able to:

- Understand the concept of energy in deforming materials.
- Define work and its relation to energy.
- Understand conservative forces and their relation to potential energy.
- Explain the concept of dissipative forces and their effect on mechanical energy.
- Understand the principle of conservation of mechanical energy.
- Describe the behavior of deforming materials under different types of forces.
- Distinguish between conservative and non-conservative internal forces.
- Apply energy methods to solve problems in mechanics.

Key Terms

- Conservation
- Conservative force
- Deformation
- Deforming materials
- Dissipative forces
- Energy
- Energy methods
- Matter
- Potential energy
- Strain
- Work

5.1. ENERGY IN DEFORMING MATERIALS

Energy comes in various forms, including mechanical, magnetic, radioactive, chemical, electrical, and others. Kinetic energies that result from the movement of an object and potential energies (possessed by a piece of matter due to a certain position) are two categories of energies (Terzopoulos et al., 1987).

A rubber ball raised above the ground possesses potential energy due to gravity. Once the ball falls, this energy gradually transforms into kinetic energy with the rise of the speed of the ball till there is only kinetic energy as the ball touches the ground. The ball starts to elastically distort when it touches the ground, gradually converting the kinetic energy into elastic strain energy kept within it. The elastic energy results from the molecules in the ball being rearranged; one can compare this to many springs that must be compressed within the ball. The ball has undergone optimum distortion all of the kinetic energy has been transformed into strain energy. The ball is again "pushed" right back for the rebound once the strain energy has been transformed into kinetic energy (Sevillano & Aernoudt, 1987) (Figure 5.2).

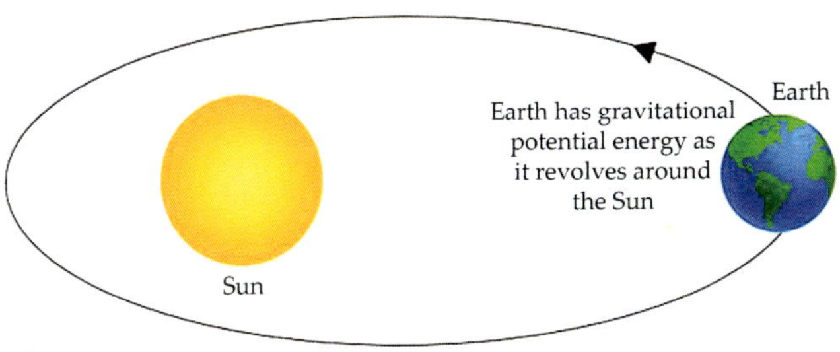

Figure 5.2. Illustration of gravitational potential energy.

Source: Sciencefacts, Creative Commons License.

In many aspects, elastic distortion of material is analogous to lifting a weight above the ground; in both situations, there is an increase in potential energy. A type of potential energy is the elastic strain energy is.

Extending a rubber band requires the same work. In rubber, this exertion is transformed into elastic strain energy. When the pulling force is gradually removed, the rubber band will "pull" back using this energy. The rubber band will swiftly return if it is extended and then immediately released; in this scenario, strain energy is transformed into sound energy ("snap") and kinetic energy (Kim & Joe, 1989). A big metal slab will experience small strains when lightweight is applied; these strains are small enough to be seen with the naked eye. Once the weight is withdrawn, the metal acts like a rubber ball and utilizes the internal strain energy to go to its original form. In comparison, a bar made of metal that has been significantly bent and set flat will not even be close to regaining its previous shape. It has been permanently altered. Most of the energy given has been wasted; it has been transformed into heat energy, causing a very small rise in the bar's temperature. This form of permanent distortion is explained by the theory of plasticity (Curtze & Kuokkala, 2010).

Any deforming natural material will partially heat up from the energy input. Nonetheless, all of the energy given is expected to be transformed into strain energy when using the perfect elastic material being studied in this chapter.

An elastic ball that falls onto a completely elastic surface will rebound up to the exact height from whence it was dropped, and the material recovers to its initial form when the loads are removed (Zhu et al., 2010).

In the next sections, some key ideas from elementary particle physics are discussed as a preliminary to a study of the energy of elastic materials. It is demonstrated that Newton's second law, the laws of kinetic energy and work, and the laws of conservation of mechanical energy are all similar claims that may be deduced from one another. After that, these ideas are applied to studying the energetics of elastic materials (Bark et al., 2021).

5.2. WORK

Consider a force F which acts on a particle, causing it to move through a displacement s, the directions in which they act being represented by the arrows in Figure 5.3(a). The work W done by F is defined to be $Fs\cos\theta$ where θ is the angle formed by positioning the start of the F and s arrows at the same location with 0–180. Work can be positive or negative: when the force and displacement are in the same direction, then 0–90 and the work done is positive; when the force and displacement are in opposite directions, then 90–180 and the work done is negative (Marlevi et al., 2019).

Consider next a particle moving along a certain path between the points p_1, and p_2 by the action of some force F. The work done is:

$$W = \int_{p_1}^{p_2} F \cos\theta \, ds \quad (5.1)$$

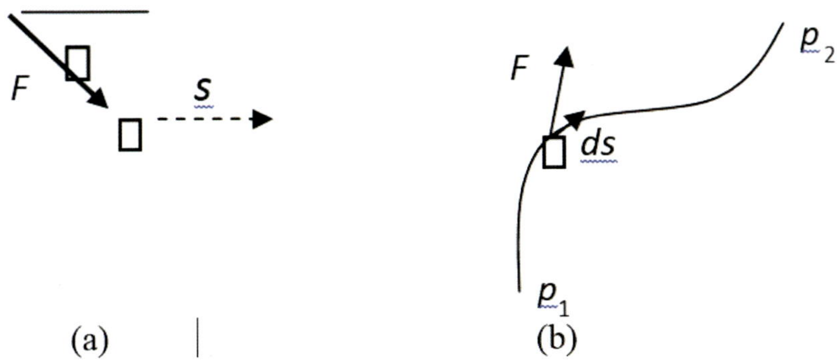

Figure 5.3. Illustration of (a) Force acting on a particle, which moves through a displacements; and (b) a varying force moving a particle along a path.

where; s is the displacement. For motion along a straight line, so that θ=0, the work is $W = \int_{p_1}^{p_2} F ds$; if F here is constant, then the work is simply F times the distance between p_1 to p_2 but, in most applications, the force will vary, and an integral need to be evaluated (Coulomb & Meunier, 1984).

5.3. CONSERVATIVE FORCES

Eqn. (5.1) states that the path taken will typically influence how often work is done by the force to drive a particle to a given distance. However, there are many crucial real-world cases where the work is not dependent on the direction but depends on the starting and finishing points, like when bending elastic materials. These circumstances lead to the idea of a conservative force (Visser, 2011). When addressing the one-dimensional case, a conservative force F_{con} could be given as the derivative of a function U,

$$F_{con} = -\frac{dU}{dx}, \quad (5.2)$$

since, in that case,

$$W = \int_{p_1}^{p_2} F_{con} \, dx = -\int_{p_1}^{p_2} \frac{dU}{dx} dx = -\int_{p_1}^{p_2} dU = -(U(p_2) - U(p_1)) = -\Delta U \quad (5.3)$$

In this case, the term "potential energy" refers to the function U, and "U" denotes the variation in potential energy of an object as it goes from p_1 to p_2. The potential energy U of the particle is the same as when it began if the object is transported from position p_1 to position p_2, then back to position p_1 (Done & Simpson, 1977).

5.4. POTENTIAL ENERGY

The potential energy of any object is given as:

The work done in driving a system from some standard configuration to the present configuration.

There are the following features of potential energy:

1. Presence of a force field;
2. Required work done to drive anything in the force field;
3. Presence of conservative force fields;
4. Some reference configurations;
5. Negative work done by a force field in the presence of opposing force;
6. Recoverable energy.

The following example demonstrates these six features: A force F steadily extends an object coupled to a spring coil, defeating the spring force F_{spr} (with no accelerations and $F = -F_{spr}$) (Lorenz, 1955).

Figure 5.4. A force extending an elastic spring.

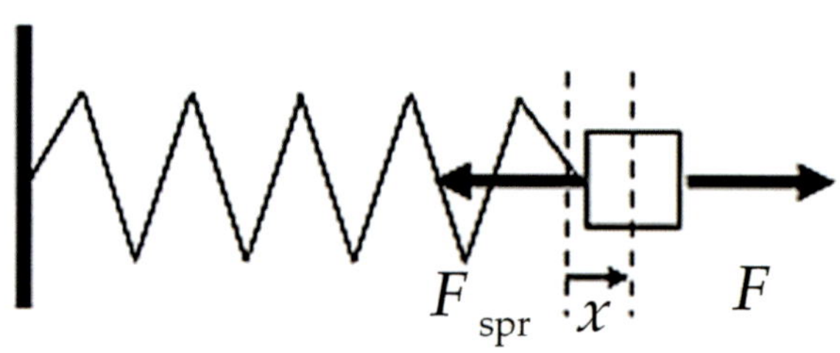

Source: Science Ready, Creative Commons License.

Consider that the block has an actual position of x_0 (x=0 as reference configuration). Imagine the force and deflection to be proportional, F =kx, the work done by F in stretching the spring to a certain distance x is given by:

$$W = \int_{x_0}^{x} F dx = \int_{x_0}^{x} kx dx = \tfrac{1}{2}kx^2 - \tfrac{1}{2}kx_0^2 \equiv U(x) - U(x_0) = \Delta U \tag{5.4}$$

This refers to the potential energy used to drive something within the "force field" of an elastic spring. Since this spring is prepared to draw back and perform the same extent of work, it is believed that the energy used to move the body can be recovered.

The resulting work done by the conservative force F_{spr} in spring is:

$$W_{spr} = -\int_{x_0}^{x} F dx = -\left(\tfrac{1}{2}kx^2 - \tfrac{1}{2}kx_0^2\right) \equiv -\Delta U \qquad (5.5)$$

The triangles area in Figure 5.5 demonstrates this work – in the reference position x=0, the spring force is zero, gradually increasing as the value of x increases (Simons et al., 1983).

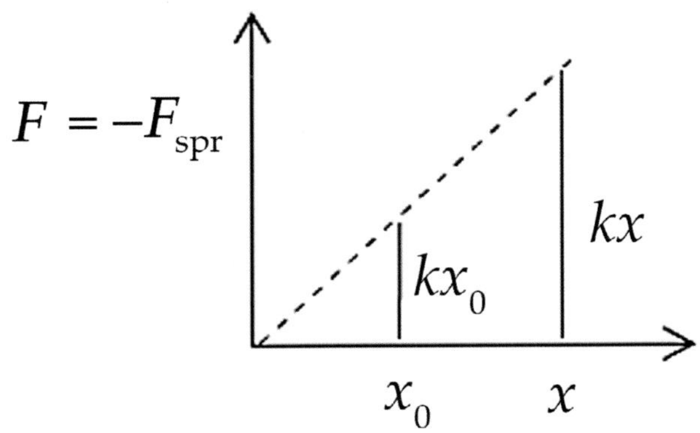

Figure 5.5. Force-extension curve for a spring.

Source: Hyperphysics, Creative Commons License.

The forces in this illustration vary according to how far the spring is extended. The potential energy will be influenced by the distance between particles of material, similar to how potential energy is retained in any material.

Furthermore, the example demonstrates that a different definition for a system's potential energy, U is the work done by a conservative force to change the protocol from a predetermined framework to the present state (Schlegel, 2003).

So, in practice, work done by conservative force and potential energy is connected through

$$W_{con} = -\Delta U \qquad (5.6)$$

5.5. DISSIPATIVE (NON-CONSERVATIVE) FORCES

One cannot discover a general function whenever the forces are dissipative instead of conservative. One must consider the particle's route because the work done will vary depending on every situation, so the work done is really the variation between the U values at the initial and final regions (Anderson, 1974). The fact that the total work done will not be zero when a particle is moved and returned to its initial place is a common characteristic of non-conservative forces. Consider Figure 5.6 to illustrate a block being pulled over a rough surface. In this example, if the block drags back and forth repeatedly, the pulling force of F's work increases gradually. The amount of work done depends on the block's ultimate location and its entire route history. The block is moved using energy, which is then lost as heat (the energy is not recoverable) (Lemeshko & Weimer, 2013).

Figure 5.6. Dragging a block over a frictional surface.

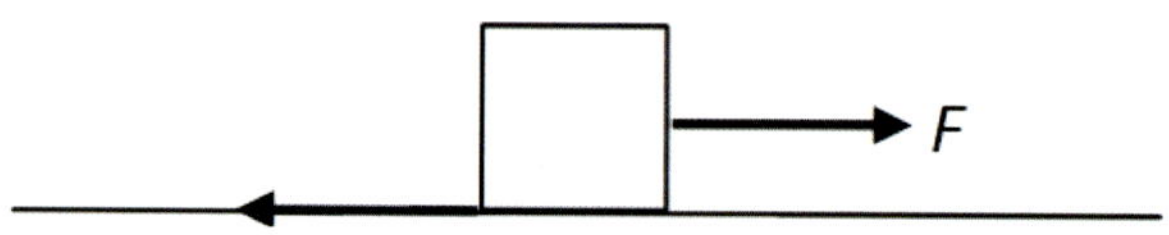

Source: Hyper Physics, Creative Commons License.

5.6. THE PRINCIPLE OF WORK AND KINETIC ENERGY

In particular, Newton's second law, or the concept of work and energy, can be used to resolve any problem in mechanics (which is discussed here). Here are two alternative equations that essentially express a similar concept, but based on the issue, one may be better. The work-energy concept defines problems in the context of energy, whereas Newton's second law addresses forces (Sears et al., 1980).

K = mv^2 is the relation for the kinetic energy of an object with velocity v and mass m. Considering Newton's second rule F = ma, the rate at which kinetic energy changes is:

$$\dot{K} = \frac{d}{dt}\left(\frac{1}{2}mv^2\right) = mv\frac{dv}{dt} = (ma)v = Fv \tag{5.7}$$

The change in kinetic energy over a time interval (t_0, t_1) is then:

$$\Delta K = K_1 - K_0 = \int_{t_0}^{t_1} \frac{dK}{dt} dt = \int_{t_0}^{t_1} Fv dt \quad (5.8)$$

where K_0 and K_1 are the initial and final kinetic energies. The work done over this time interval is

$$W = \int_{W(t_0)}^{W(t_1)} dW = \int_{x(t_0)}^{x(t_1)} F dx = \int_{t_0}^{t_1} Fv dt \quad (5.9)$$

and it follows that

$$\boxed{W = \Delta K} \quad (5.10)$$

The principle of work and kinetic energy:

The total work done by the external forces acting on a particle equals the change in kinetic energy of the particle.

It is not a new principle of mechanics, rather a rearrangement of Newton's second law of motion (or one could have started with this principle, and derived Newton's second law) (Peñate et al., 2010).

Remember

The principle of work and kinetic energy states that the work done on an object is equal to the change in its kinetic energy, which is useful in solving problems related to the motion of objects where the net work done is known but the force acting is not.

5.7. THE PRINCIPLE OF CONSERVATION OF MECHANICAL ENERGY

As an assumption, there is no loss of energy and no action by dissipative forces. A body's potential and kinetic energies are combined to form the complete mechanical energy of the body. In this particular example, the work-energy concept might be stated in one of two ways:

1. In a body, the total work done by external forces is equivalent to the difference in kinetic energy of the body:

$$W = W_{con} + W_{apl} = \Delta K \quad (5.11)$$

2. In a body, the total work done by external forces, specially the conservative forces, is equivalent to the alteration in the total mechanical energy of the body

$$W_{apl} = \Delta U + \Delta K \qquad (5.12)$$

As a result, the mechanical energy is constant in a particular situation with no external forces or all conservative external forces (Karapetkov et al., 2019).

Energy can never be generated or destroyed; but it can just be converted from one form to another, in accordance with the principle of energy conservation.

The principle of mechanical energy conservation, also known as the principle of energy conservation in the absence of energy dissipation, specifies that when a system is only under the influence of conservative forces, the mechanical energy will remain constant; any system under the control of non-conservative forces will invariably require non-mechanical energy (Bryan, 2010).

So, for the action of only conservative forces,

$$0 = \Delta U + \Delta K \qquad (5.13)$$

or,

$$K_f + U_f = K_i + U_i \qquad (5.14)$$

Here K_f and K_i denotes the kinetic energies (final and initial), and U_f and U_i denotes potential energies (final and initial).

It should be noted that the principle of work-energy (also known as Newton's second law) is re-expressed as the concept of mechanical energy conservation (Fahrunnisa et al., 2021).

TIPS

When designing with deforming materials, consider their non-linear behavior like strain hardening where stress required to continue deforming increases. Use specialized testing and modeling techniques to account for this behavior and avoid unexpected failure modes.

5.8. DEFORMING MATERIALS

The concept of the mechanics of particles has now been expanded to include a deforming substance.

Every substance comprises several particles that interact in various intricate ways. When the material is in a natural equilibrium condition, a complex network of intrinsic forces will be at work between the molecules. When external forces are exerted, the material will distort, also, molecules will consider moving; therefore, the work will be performed by both internal and external forces (Benn & Evans, 1996).

According to the work-energy principle, all work performed by external and internal forces in this situation is equivalent to the difference in kinetic energy.

$$W_{ext} + W_{int} = \Delta K \quad (5.15)$$

For a system with no external forces, one has

$$W_{int} = \Delta K \quad (5.16)$$

This situation is referred to as free vibration. The situation with the constant kinetic energy is

$$W_{ext} + W_{int} = 0 \quad (5.17)$$

This case is quasi-static (the values still time dependent).

The attractive forces among the molecules must be categorized:

- internal forces of conservative systems;
- the internal force of non-conservative systems (Wolf et al., 2005).

5.9. CONSERVATIVE INTERNAL FORCES

Firstly, one can consider that the molecules contact one another as elastic springs, assuming a system of conservative internal force. As seen in Figure 5.7, consider using an external force to separate two of these molecules.

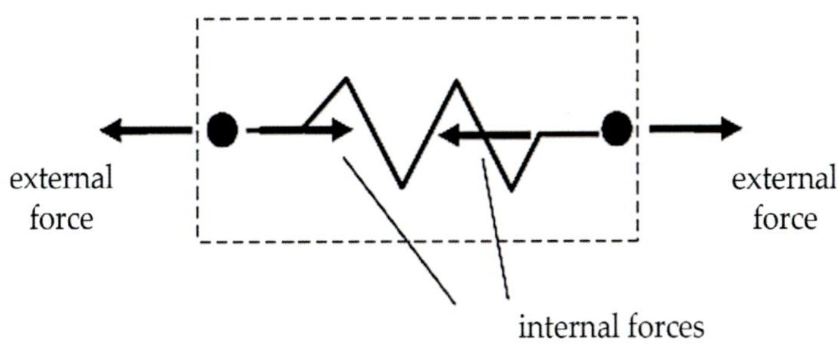

Figure 5.7. External force pulling two molecules/particles apart.

Source: Science Facts, Creative Commons License.

In this perfect situation, it is possible to state that the work performed by external forces is the sum of the changes in and kinetic and potential energy.

$$W_{ext} = \Delta U + \Delta K \quad (5.18)$$

Did you Know?

Conservative internal forces, such as gravity and spring forces, are path-independent, meaning that the work done by these forces on an object is only dependent on its initial and final positions, making them easier to calculate using potential energy functions.

The energy U in this condition of distorting materials is known as the elastic strain energy, the energy produced by the symmetry of molecules similar to some equilibrium position.

The free vibration is $0 = \Delta U + \Delta K$, and quasi-static is $W_{ext} = \Delta U$ (Widnall & Peraire, 2008).

5.10. NON-CONSERVATIVE INTERNAL FORCES

Suppose a polymer with molecules arranged in long-chain is another illustration of internal forces at work in materials. These molecules would pass over one another if an external force were applied to a pair of molecules, as depicted in Figure 5.7. As the frictional force produced between the block and the rough surface in Figure 5.8, frictional forces would exist among the molecules. Internal friction is the term for this. As the work relies on the selected path, even if the internal forces are dissipative, it is impossible to express the external work regarding potential energy $W_{ext} \neq \Delta U + \Delta K$. The amount of work would need to be calculated by calculating an integral (Winkler et al., 2019).

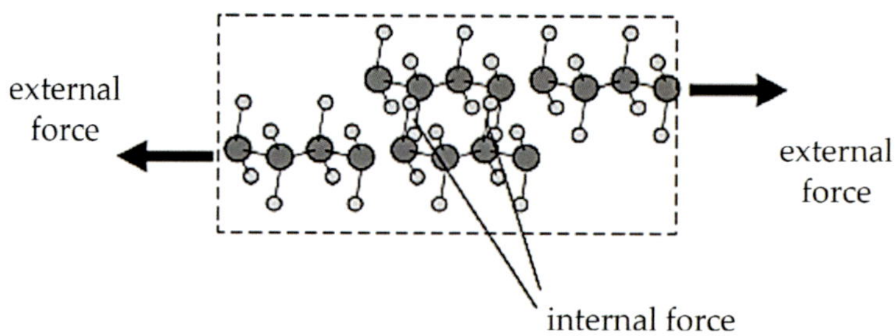

Figure 5.8. An external force separating two molecules by a pulling force.

Like Eqn. (5.12), the energy balance can be given as

$$W_{ext} - H = \Delta K + \Delta U \quad (5.19)$$

According to the specific deformation protocol, H is the energy lost during distortion. Heat transfer causes this energy to evaporate, then be transported away through the substance (Kirillov & Seyranian, 2005).

5.11. ENERGY METHODS

The energy protocol is an effective way to analyze structural components to ascertain their deflection, stability, and other characteristics. It is a technique that depends on energy conservation and is predicated on the idea that a system's total energy remains constant. According to this approach, the energy of a building is determined by taking into account both the work that has been done on it and the potential energy it contains. The structure's behavior is then calculated by equating this energy to the external work done on it. Plates, frames, and beams are just a few structures that can be used with the energy approach. The energy approach and its usage in a basic beam structure are described below (Antontsev et al., 2002).

Imagine a beam of length L being carried with a constant amount of weight w. The assumption is that the material of the beam is isotropic and homogenous with a cross-sectional area (A) and elastic modulus (E). Y(x), where; x is the length of the beam, stands for the deviation of the beam.

The beam's potential energy and kinetic energy can be combined to form its overall energy. Due to the low velocity of a beam, its kinetic energy is insignificant, and the elastic distortion of the beam gives it its potential energy. The work performed by external forces on the beam to distort it from its initial condition to its deformed state is used to determine this potential energy (Solans & Solé, 2012).

The equation gives the potential energy of the beam:

$$U = \int (1/2)\, EI\, (y'')^{2} dx - \int wy dx$$

Here the first integral reflects the stored elastic energy in the beam owing to bending, and the second integral reflects the work done by the external force. E denotes the modulus of elasticity. I denote the moment of inertia of the beam.

According to energy conservation, the system's overall energy is constant. As a result, potential energy U is equivalent to the external work done on it. Calculating the external work done on a beam involves multiplying the load w by the movement of the beam at its open end (Straughan, 2013).

$$W = \int wy dx = PL/2$$

Here P denotes the complete load on the beam.

The work done on the beam W and potential energy U gives an equation:

$$U = W$$

or

$\int(1/2)EI(y'')^2dx - \int wydx = PL/2$

This equation calculates the distortion in a beam, which is also termed the principle of virtual work. The deflection equation may be determined by reducing the total potential energy U concerning the deflection y(x). The partial derivation of U with regard to y(x) is taken in order to achieve this, and it is drawn to zero.

$dU/dy(x) = 0$

It gives a differential equation:

$EIy'''' = wy$

where; y'''' is the fourth derivative of y(x).

There are several ways to solve this equation, like finite element method, superposition method, and moment-area approach. The answer gives the beam's deflection y(x) under the specified loading situation.

In short, an energy approach is a potent tool for structural analysis. It is predicated on the idea that a system's total amount of energy is constant and depends on the principle of energy. The technique can be used on many different structures of various types, and the deflection equation could be found by reducing the structure's overall potential energy. The energy technique, frequently used in engineering, offers an easy and simple strategy for resolving challenging structural issues (Lyubartsev et al., 1992).

ACTIVITY 5.1.

A spring-loaded mechanism has a mass of 50 kg attached to it. The spring is initially compressed by 20 cm and released, causing the mass to move forward until it reaches a stopper. If the spring has a stiffness of 500 N/m, calculate the velocity of the mass just before it hits the stopper using the principle of virtual work and energy. Assume there is no friction or air resistance.

5.12. SUMMARY

The chapter on Virtual Work and Energy covers the concepts of work, energy, and forces in mechanics. It starts with the concept of energy in deforming materials and explains how work is related to energy. The chapter then covers conservative forces and their relation to potential energy, as well as the concept of dissipative forces and their effect on mechanical energy. The principle of conservation of mechanical energy is introduced, and the behavior of deforming materials under different types of forces is discussed. The chapter also distinguishes between conservative and non-conservative internal forces and applies energy methods to solve problems in mechanics.

REVIEW QUESTIONS

1. What is the relationship between work and energy?
2. Explain the concept of conservative and non-conservative forces.
3. Describe the principle of conservation of mechanical energy.
4. What are dissipative forces, and how do they affect mechanical energy?
5. How are energy methods used to solve problems in mechanics?

MULTIPLE CHOICE QUESTIONS

1. What is work?

a. A force acting on an object
b. The product of force and time
c. The product of force and displacement
d. The product of mass and acceleration

2. What is a conservative force?

a. A force that depends on the velocity of an object
b. A force that depends on the position of an object
c. A force that does not depend on the path taken by an object
d. A force that is always perpendicular to the motion of an object

3. What is potential energy?

a. Energy that is stored in a system due to its position or configuration
b. Energy that is lost due to friction
c. Energy that is transferred from one object to another
d. Energy that is produced by a force acting on an object

4. What are dissipative forces?

a. Forces that increase the mechanical energy of a system

b. Forces that do not affect the mechanical energy of a system
c. Forces that decrease the mechanical energy of a system
d. Forces that change the direction of motion of an object

5. **What is the principle of conservation of mechanical energy?**
   a. The total energy of a system is conserved if no external forces act on it.
   b. The total momentum of a system is conserved if no external forces act on it.
   c. The total angular momentum of a system is conserved if no external torques act on it.
   d. The total work done on a system is equal to the change in its kinetic energy.

Answers to Multiple Choice Questions

1. (c) 2. (c); 3. (a) 4. (c); 5. (a)

REFERENCES

1. Anderson, G. L., (1974). Optimal design of a cantilever subjected to dissipative and non-conservative forces. Journal of Sound and Vibration, 33(2), 155–169.
2. Antontsev, S. N., Díaz, J. I., Shmarev, S., & Kassab, A. J. (2002). Energy Methods for Free Boundary Problems: Applications to Nonlinear PDEs and Fluid Mechanics. Progress in Nonlinear Differential Equations and Their Applications, 55(4), B74-B75.
3. Bark, H., Tan, M. W. M., Thangavel, G., & Lee, P. S., (2021). Deformable high loading liquid metal nanoparticles composites for thermal energy management. Advanced Energy Materials, 11(35), 1–50.
4. Benn, D. I., & Evans, D. J., (1996). The interpretation and classification of subglacially-deformed materials. Quaternary Science Reviews, 15(1), 23–52.
5. Bryan, J. A., (2010). Investigating the conservation of mechanical energy using video analysis: Four cases. Physics Education, 45(1), 50.
6. Coulomb, J. L., & Meunier, G., (1984). Finite element implementation of virtual work principle for magnetic or electric force and torque computation. IEEE Transactions on Magnetics, 20(5), 1894–1896.
7. Curtze, S., & Kuokkala, V. T., (2010). Dependence of tensile deformation behavior of TWIP steels on stacking fault energy, temperature and strain rate. Acta Materialia, 58(15), 5129–5141.
8. Done, G. T. S., & Simpson, A., (1977). Dynamic instability of certain conservative and non-conservative systems. Journal of Mechanical Engineering Science, 19(6), 251–263.
9. Halilovic, A., Mešic, V., Hasovic, E., & Dervic, D., (2021). Students' difficulties in applying the law of conservation of mechanical energy: Results of a survey research. European Educational Researcher, 4(2), 171–192.

10. Kadayif, I., Chinoda, T., Kandemir, M., Vijaykirsnan, N., Irwin, M. J., & Sivasubramaniam, A., (2001). vEC: Virtual energy counters. In: Proceedings of the 2001 ACM SIGPLAN-SIGSOFT Workshop on Program Analysis for Software Tools and Engineering (Vol. 1, pp. 28–31).
11. Karapetkov, S., Dimitrov, L., Uzunov, H., & Dechkova, S., (2019). Identifying vehicle and collision impact by applying the principle of conservation of mechanical energy. Transport and Telecommunication Journal, 20(3), 191–204.
12. Kim, B. H., & Joe, C. R., (1989). Single specimen test method for determining fracture energy (J_c) of highly deformable materials. Engineering Fracture Mechanics, 32(1), 155–161.
13. Kirillov, O. N., & Seyranian, A. O., (2005). The effect of small internal and external damping on the stability of distributed non-conservative systems. Journal of Applied Mathematics and Mechanics, 69(4), 529–552.
14. Lemeshko, M., & Weimer, H., (2013). Dissipative binding of atoms by non-conservative forces. Nature Communications, 4(1), 2230.
15. Li, H., Giles, W., Ren, L., Bowsher, J., & Yin, F. F., (2012). Implementation of dual-energy technique for virtual monochromatic and linearly mixed CBCTs. Medical Physics, 39(10), 6056–6064.
16. Lorenz, E. N., (1955). Available potential energy and the maintenance of the general circulation. Tellus, 7(2), 157–167.
17. Lyubartsev, A. P., Martsinovski, A. A., Shevkunov, S. V., & Vorontsov Velyaminov, P. N., (1992). New approach to Monte Carlo calculation of the free energy: Method of expanded ensembles. The Journal of Chemical Physics, 96(3), 1776–1783.
18. Marlevi, D., Ruijsink, B., Balmus, M., Dillon-Murphy, D., Fovargue, D., Pushparajah, K., & Nordsletten, D. A., (2019). Estimation of cardiovascular relative pressure using virtual work-energy. Scientific Reports, 9(1), 1–16.
19. Peñate, B., De La Fuente, J. A., & Barreto, M., (2010). Operation of the RO Kinetic® energy recovery system: Description and real experiences. Desalination, 252(1–3), 179–185.
20. Schlegel, H. B., (2003). Exploring potential energy surfaces for chemical reactions: An overview of some practical methods. Journal of Computational Chemistry, 24(12), 1514–1527.
21. Sears, S. B., Parr, R. G., & Dinur, U., (1980). On the quantum mechanical kinetic energy as a measure of the information in a distribution. Israel Journal of Chemistry, 19(1–4), 165–173.
22. Sevillano, J. G., & Aernoudt, E., (1987). Low energy dislocation structures in highly deformed materials. Materials Science and Engineering, 86, 35–51.
23. Simons, J., Joergensen, P., Taylor, H., & Ozment, J., (1983). Walking on potential energy surfaces. The Journal of Physical Chemistry, 87(15), 2745–2753.

24. Solans, C., & Solé, I., (2012). Nano-emulsions: Formation by low-energy methods. Current Opinion in Colloid & Interface Science, 17(5), 246–254.
25. Terzopoulos, D., Platt, J., Barr, A., & Fleischer, K., (1987). Elastically deformable models. In: Proceedings of the 14th Annual Conference on Computer Graphics and Interactive Techniques (pp. 205–214).
26. Visser, M., (2011). Conservative entropic forces. Journal of High Energy Physics, 2011(10), 1–22.
27. Winkler, R., Kindmann, R., & Knobloch, M., (2019). Assessment of the plastic capacity of I-shaped cross-sections according to the partial internal forces method. Engineering Structures, 191, 740–751.
28. Wolf, D., Yamakov, V., Phillpot, S. R., Mukherjee, A., & Gleiter, H., (2005). Deformation of nanocrystalline materials by molecular-dynamics simulation: Relationship to experiments? Acta Materialia, 53(1), 1–40.
29. Zhu, F., Lu, G., Ruan, D., & Wang, Z., (2010). Plastic deformation, failure and energy absorption of sandwich structures with metallic cellular cores. International Journal of Protective Structures, 1(4), 507–541.
30. Zuhe, S., & Wolfe, M. A., (1990). On the existence of periodic solutions of periodically perturbed conservative systems. Journal of Mathematical Analysis and Applications, 153(1), 78–83.

CHAPTER 6

MECHANICAL VIBRATIONS

UNIT INTRODUCTION

Mechanical vibration is a subfield of engineering mechanics that studies oscillatory motion inside mechanical components. Vibrations are pervasive in practically all mechanical systems, including buildings and bridges to automobiles and equipment, making it a vital engineering research subject. Vibrations could induce mechanical system stress, wear and tear, and even catastrophic failure, causing major repercussions (Rao & Yap, 1995).

Mechanical vibration research entails knowing basic vibration principles, including frequency, phase, and amplitude. It also entails analyzing other vibrations, including forced and free vibrations of multiple degrees of freedom systems. It also discusses various vibration control approaches, like active and passive controls (Findeisen & Popp, 2002). The reason for investigating the topic of mechanical motion is to gain an understanding of the behavior of mechanical components in a variety of vibration situations, to foresee and avert issues that are caused by vibration, and to design and maximize mechanical systems to reduce or do away with vibration as much as possible. It necessitates an in-depth comprehension of the physical and mathematical concepts that govern vibrations and an in-depth grasp of the methodologies involved in vibration assessment and management (Parker et al., 1990).

In general, the analysis of mechanical motion is an important subfield of engineering that plays an important part in guaranteeing mechanical systems' safety, dependability, and performance when used in various applications. Mechanical motion analysis is an example of interdisciplinary research since it draws on concepts from various disciplines, including mechanics, physics, mathematics, and materials science (Varanis et al., 2018) (Figure 6.1).

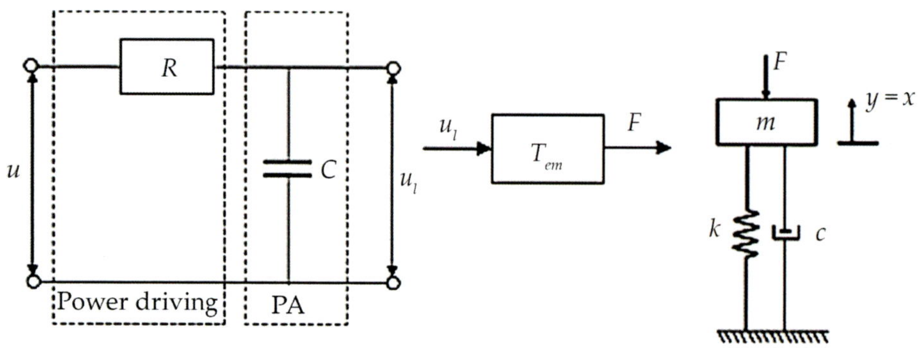

Figure 6.1. Illustration of mechanical vibrations.

Source: Science Direct, Creative Commons License.

Learning Objectives

At the end of this chapter, reader will be able to understand:

- The mechanism of simple harmonic motion;
- The concept of single degree of freedom system;
- The process of free vibration without damping and with viscous damping;
- The concept of forced vibration of two degree and multi-degree of freedom;
- The significance of vibration control and vibration suppression;
- The applications of vibrations.

Key Terms

- Active vibration control
- Atmospheric turbulence
- Damped frequency
- Natural frequency
- Orthogonality of modes
- Passive vibration control
- Simple harmonic motion
- Simple vibratory motion
- Spectral matrices

6.1. SIMPLE HARMONIC MOTION

In physics, "simple harmonic motion" refers to a movement repeated through an equilibrium, or center, position so that the greatest displacement on either side of this location is equivalent to the greatest displacement on the opposing side. The driving force, which is always oriented toward the location of equilibrium and also is directly proportionate to the distance from it, is the one that is accountable for the motion. $F = -kx$, where F represents the force, x represents the displacement, and k represents a constant value. Hooke's law is the name given to this connection (Galeriu et al., 2014) (Figure 6.2).

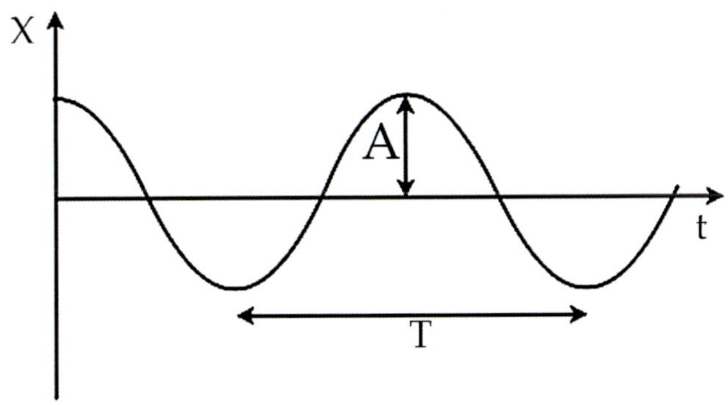

Figure 6.2. Graphical representation of simple harmonic motion.

Source: University of Birmingham, Creative Commons License.

The oscillation of a mass attached to either a vertical spring or has the opposite end of a spring anchored in ceilings illustrates a specific type of simple harmonic oscillator. At the point of maximum displacement, x, the spring is stretched to its limit, exerting the most upward force possible on the mass. At the point of maximum displacement +x, the spring will have reached its peak compression, which will cause the mass to move back in the opposite direction. At either location of maximum displacement, its force becomes the largest and is directed toward the equilibrium condition (Somroob & Wattanakasiwich, 2017). At the same time, the mass's velocity (v) is at its lowest point, its acceleration is at its highest point, and the mass moves in the opposite direction. When we are in equilibrium, our velocity is at its highest point, and acceleration (a) has stopped completely. This changing acceleration which is constantly directed towards the equilibrium point and is correlated with the movement from the equilibrium point is one of the defining characteristics of simple harmonic motion. The above acceleration is proportional to a displacement from the equilibrium point. In addition, the duration of the gap between each full vibration is unaffected by the magnitude of the greatest displacement and remains constant throughout the process. As such, in some shape or form, the most fundamental aspect of keeping time involves simple harmonic motion (Triana & Fajardo, 2011).

To demonstrate how the mass's displacement evolves, one could use Newton's second law, F = ma, and set:

$$ma = -kx$$

The resulting differential equation may be solved using the acceleration, i.e., is the second derivative of x about the time t.

$$x = A \cos \omega t,$$

Remember

The amount of time that elapses between each full vibration stays the same.

where; is the ω angular frequency expressed in radians every second; and A is the largest displacement. The amount of time required for the mass to travel from point A -to point A and back again for ωt to advance by 2π. Therefore, the time T required for the mass to travel from point A to point A and back is:

$$\omega_T = 2\pi, \text{ or } T = 2\pi/\omega.$$

The vibration's frequency, expressed in cycles per second, is 1/T or ω/2π (Kinchin, 2016). Numerous physical systems, such as an oscillating pendulum, electrons in a wire carrying alternating current, vibrating moderate particles inside a sound wave, or other assemblages involving comparatively small oscillations about such a situation of stable equilibrium, display simple harmonic motion (assuming no energy loss) (Lan & Shih, 2012). Since musical instruments create vibrations that match sound waves traveling through the air, the motion is known as harmonic. A musical instrument's vibrating components oscillate in various ways, represented by groups of superimposed simple harmonic movements, each of which frequencies is a multiple of the minimum fundamental frequency. These motions are heard as musical notes. The French mathematician Joseph Fourier discovered that every regularly recurring motion or wave, regardless of how complex its shape, may be thought to be the sum of a collection of precise harmonic movements or waves (Lan, et al., 2012).

6.2. SINGLE DEGREE OF FREEDOM SYSTEMS (SDOF)

A single coordinate is needed to fully define the setup of a single degree of freedom (SDOF) system (For the time being, this term works well.) While several options are often available, some are more logical than others for the coordinates to be utilized. We have a single governing differential motion equation for a system including one degree of freedom (DOF). The precise nature of the issue determines the details of the equation (Christopoulos et al., 2003).

We start our investigation into vibrations by considering a system's free vibrations; understanding such systems is essential to understand more complex subjects in vibration since they are the simplest to deal with. We start by analyzing undamped free vibrations to make things as straightforward as possible. In later sections, the impacts of damping will be incorporated (Nayfeh, 1983).

In a typical free vibration case, a mass is moved out of balance but released at either rest or a particular starting velocity. The mass would frequently shift back towards its equilibrium location because of the restoring forces within the system, but it will exceed the equilibrium position because of its inertia. The mass would eventually come to a temporary stop and begin to move back toward its equilibrium point, and the cycle would continue (indefinitely if there is no damping). In this case, we aim to identify and comprehend the mathematics employed to describe that motion (Nayfeh, 1985) (Figure 6.3).

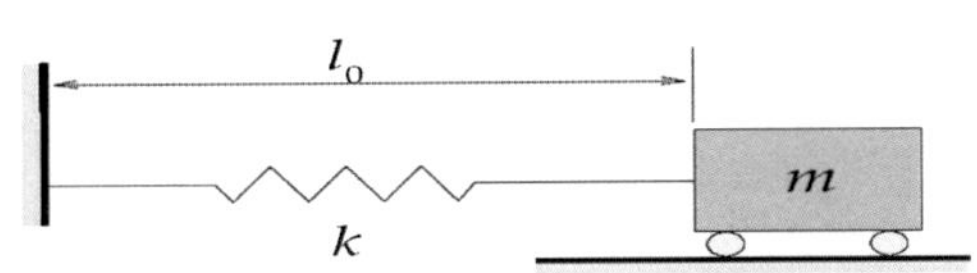

Figure 6.3. Diagram of simple vibratory motion.

Source: Engcourses, Creative Commons License.

Observe Figure 6.3, perhaps the most basic vibratory system imaginable. Here, a spring having constant stiffness k holds a mass m body to a rigid support. Its spring's length, when not extended, is low. Since we only require one coordinate (the position of a mass) to characterize the system's design fully, it is an SDOF system. We have many coordinate options (Basili et al., 2019) (Figure 6.4).

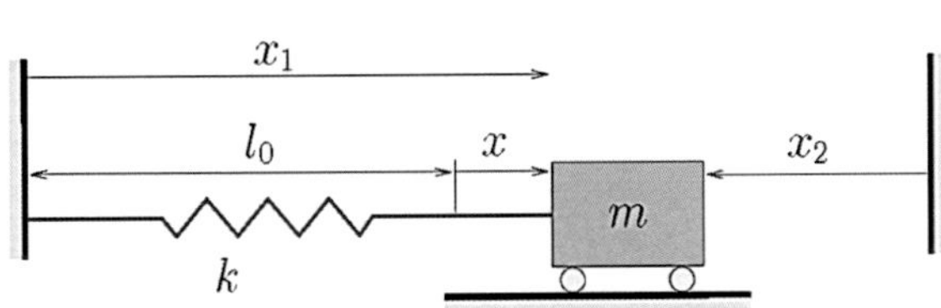

Figure 6.4. Image of several possible coordinates for simple SDOF system.

Source: Engcourses, Creative Commons License.

Three such coordinates are shown in Figure 6.4, although clearly, many more might be selected. In this diagram, although x_1 and x_2 are measured from various fixed positions, x is measured from the equilibrium point. All these options for the coordinates are legitimate, but as we shall see from our research on vibrations, there are benefits to measuring displacements first from the equilibrium point, so this is what we will do in this case (Marino & Cicirello, 2020).

We shall use Newton's Laws to determine the equation of motion. To achieve this, we consider how far the system has moved in the direction of the positive coordinates and create an FBD/MAD of a system in this situation. We also suppose that at this point, the system's acceleration and velocity are positive (i.e., only in a positive coordinate direction). This process will ensure that all the terms in your governing equations have the appropriate and consistent signs. Keep in mind that such selection of the coordinate determines the positive coordinate direction. The positive direction in Figure 6.4 is to the right for x and x_1 and left for x_2. Everything is acceptable and will function, but once you've chosen that coordinate (and consequently the direction of the positive coordinates), you should stick with it throughout the rest of the process (Leardini et al., 1999) (Figure 6.5).

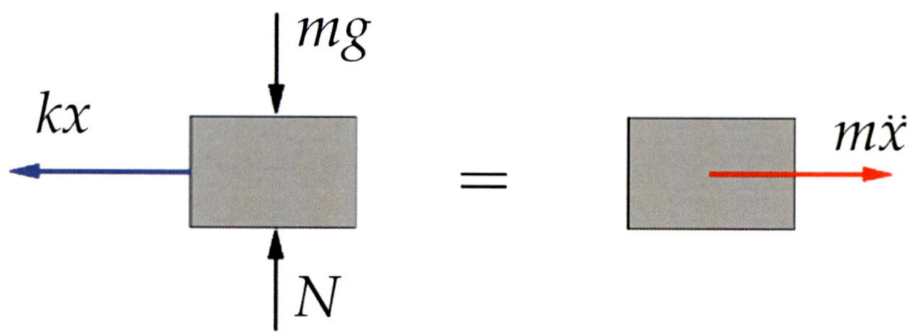

Figure 6.5. FBD/MAD of mass using as the coordinate (assuming $x, \ddot{x} > 0$).

Source: Science Direct, Creative Commons License.

Figure 6.5 uses as the coordinate and displays the FBD/MAD for our system. Using Newton's Laws, we get at (Udwadia & Tabaie, 1981).

$$\xrightarrow{+} \sum F = ma: \quad -kx = m\ddot{x}$$

$$m\ddot{x} + kx = 0 \tag{6.1}$$

Dividing by m results in:

$$\ddot{x} + \frac{k}{m}x = 0$$

and defining

$$p^2 = \frac{k}{m} \quad (6.2)$$

Allows Eqn. (6.1) to be written as:

$$\ddot{x} + p^2 x = 0 \qquad (6.3)$$

Eqn. (6.3) is the equation of motion's standard form for the SDOF system's undamped free vibration analysis. Another coordinate could replace x in Eqn. (6.3) if that were utilized. As an illustration, if we had picked the coordinate x_2, the motion formula would take the form (Baltzopoulos et al., 2018).

$$\ddot{x}_2 + p^2 x_2 = 0$$

Eqn. (2.3) is a constant coefficient second-order linear ordinary differential equation (ODE). Hence, we consider a solution of the following type:

$$x(t) = Ce^{st} \qquad (6.4)$$

This, when added to Eqn. (6.3), yields (Cosenza et al., 2009)

$$\left[\cancel{C}s^2\cancel{e^{st}}\right] + p^2\left[\cancel{Ce^{st}}\right] = 0$$

$$s^2 + p^2 = 0$$

Which have two roots given?

$$s = \pm ip$$

As a result, the answer to Eqn. (6.3) is.

$$x(t) = C_1e^{ipt} + C_2e^{-ipt} \quad (6.5)$$

(C_1 and C_2 must be permitted to be complex values in this case.) To proceed, we employ Euler's identity)

$$e^{ipt} = \cos pt + i\sin pt$$

$$e^{-ipt} = \cos pt - i\sin pt$$

to write

$$x(t) = C_1[\cos pt + i\sin pt] + C_2[\cos pt - i\sin pt]$$
$$= [C_1 + C_2]\cos pt + [C_1 - C_2]i\sin pt$$

Defining*

$A=[C_1-C_2]i$

$B=[C_1+C_2]$

The solution in Eqn. (6.5) can be written as:

$$x(t) = A\sin pt + B\cos pt \quad (6.6)$$

Eqn. (6.6) is the free un-damped vibration response of an SDOF system and is the basic solution to Eqn. (6.3). The reaction is a simple harmonic motion that happens at a frequency known as the system's natural frequency. Formula 6.2 shows one for the basic spring-mass system under consideration (Cosenza et al., 2009).

$$p = \sqrt{\frac{k}{m}}$$

Remember that the typical way to represent the equation of motion for a generic SDOF system is.

$$\ddot{x} + p^2 x = 0$$

If an equation of motion is stated in this standard form, the natural frequency, which is a fundamental feature of a vibrating system, may be established by observation. Note that p in Eqn. (6.3) is always in units of rad/s (Habib et al., 2017).

In Eqn. (6.6), there must be two unknown variables, A and B. Usually; we calculate their values by examining the system's starting circumstances. We need to declare two beginning conditions because there are two unknown constants. All of those are generally the mass's initial location and velocity. For instance, at time t=0, we may have

$$x(0) = x_0$$
$$\dot{x}(0) = v_0$$

where; x_0 or v_0 might be negative.

Eqn. (6.6) then gives:

$$x(0) = x_0 = A\sin(p \cdot 0)^{0} + B\cos(p \cdot 0)^{1}$$

So that

B=x_0

Differentiating Eqn. (6.6)

$$\dot{x} = Ap\cos pt - Bp\sin pt$$

and applying the second initial state results in

$$\dot{x}(0) = v_0 = Ap\cos(p \cdot 0)^{1} - Bp\sin(p \cdot 0)^{0}$$

Or,

$$A = \frac{v_0}{p}$$

As a result, the system's answer is supplied by.

$$x(t) = \frac{v_0}{p}\sin pt + x_0\cos pt \qquad (6.7)$$

This thoroughly always determines the system's reaction $t > 0$.

6.3. FREE VIBRATION WITHOUT DAMPING

Newton's equation is given for mass m after first considering the un-damped systems of free vibrations. The force m$\ddot{x}$ that the mass applies to the springs is equal to and in the opposite direction to a force kx that the spring applies to the mass (Reid, 1956).

$$m\ddot{x} + kx = 0 \qquad (6.8)$$

where; x = 0 determines the mass's equilibrium position. The solution of Eqn. (6.8) is:

$$x = A\sin\sqrt{\frac{k}{m}}\,t + B\cos\sqrt{\frac{k}{m}}\,t \qquad (6.9)$$

where the term $\sqrt{k/m}$ is the natural angular frequency defined by:

$$\omega_n = \sqrt{\frac{k}{m}} \quad \text{rad/sec} \qquad (6.10)$$

The mass oscillates sinusoidally, and cycles are repeated constantly; the period is the length of a cycle (Feldman, 1997).

$$\tau = \frac{2\pi}{\omega_n} \quad (6.11)$$

Reciprocal of the period is the natural frequency:

$$f_n = \frac{1}{\tau} = \frac{\omega_n}{2\pi} = \frac{1}{2\pi}\sqrt{\frac{k}{m}} = \frac{1}{2\pi}\sqrt{\frac{kg}{W}} \quad (6.12)$$

where; W is the mg is the rigid body's weight that makes up the mass of a system in Figure 6.4. The solid lines in Figure 6.5 represent the relationships of Eqn. (6.8) (Cheng & Wang, 2003).

Initial Conditions: In Eqn. (6.9), B is the value of x at time t = 0, and A is equal to $\dot{x}/\omega_n$ at time t = 0. Thus, at zero time, the velocity and displacement circumstances dictate the ensuing oscillation.

1. **Phase Angle:** Eqn. (6.9) It is possible to write for the movement in oscillatory motion by inserting the frequency relation of Eqn. (6.10),

   $x = A \sin \omega_n t + B \cos \omega_n t = C \sin (\omega_n t + \theta)$ (6.13)

   where; $C = (A^2 + B^2)^{1/2}$; and $\theta = \tan^{-1} (B/A)$. The angle θ is called the phase angle.
2. **Static Deflection:** A basic mass-spring system's static deflection is caused by the mass's gravitational pull-on spring k., δ_{st} = mg/k. Substituting this relation in Eqn. (6.8).

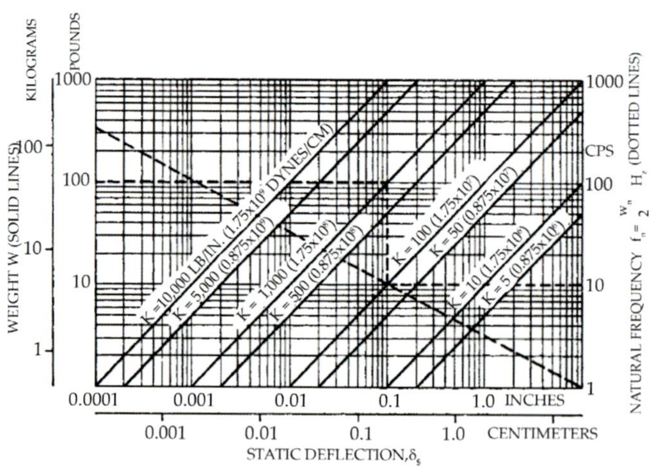

Figure 6.6. Natural frequency relationships for a system with a single degree of freedom. Solid lines in Eqn. (2.8) represent the relationship between natural

frequency, the weight of the supported body, and the spring stiffness. The diagonal-dashed line in Eqn. (2.10) illustrates the relationship between natural frequency and static deflection. The dashed line should be followed horizontally to the solid line at k = 1,000, then vertically down to the diagonal-dashed line and finally horizontally to read fn = 10 Hz from the right ordinate scale. This example shows how to find the natural frequency of a system with W = 100 lb and k = 1,000 lb/in.

Source: Bapat & Sankar, Creative Commons License.

$$f_n = \frac{1}{2\pi}\sqrt{\frac{g}{\delta_{st}}} \qquad (6.13)$$

The diagonal dashed line in Figure 6.6 represents the relationship of Eqn. (6.13). This relation applies if the system under investigation is linear and elastic. Eqn. (6.13) is irrelevant because rubber springs, for instance, frequently display a dynamic stiffness that varies from the static stiffness.

6.4. FREE VIBRATION WITH VISCOUS DAMPING

Figure 6.7 shows a system with such a viscous damper or one freedom level. Formula (6.14) represents the differential movement equation for mass m for an open-loop system (Liang, 2005).

$$m\ddot{x} + c\dot{x} + kx = 0 \qquad (2.11) \qquad (6.14)$$

Shows a system with such a viscous damper or one freedom level. Formula (2.8) represents the differential movement equation having mass m for an open-loop system (Liang & Feeny, 1998).

$$c_c = 2\sqrt{km} = 2m\omega_n \qquad (6.15)$$

The ratio $\zeta = c/c_c$ is defined as the element of critical damping.

1. Less-Than-Critical Damping: If the system's damping is less than critical, $\zeta < 1$, then the solution of Eqn. (6.14) is.

$$x = e^{-ct/2m}(A \sin \omega_d t + B \cos \omega_d t)$$
$$= Ce^{-ct/2m} \sin(\omega_d t + \theta) \qquad (6.16)$$

Concerning Eq., C and are defined Eqn. (6.13). Eqn. (6.10) establishes a relationship between the damped and undamped natural frequencies (Jafari & Azhari, 2021).

$$\omega_d = \omega_n(1 - \zeta^2)^{1/2} \qquad \text{rad/sec} \tag{6.17}$$

Eqn. (6.17), Plotting the relationship between the undamped and damped natural frequencies in (Figure 6.7):

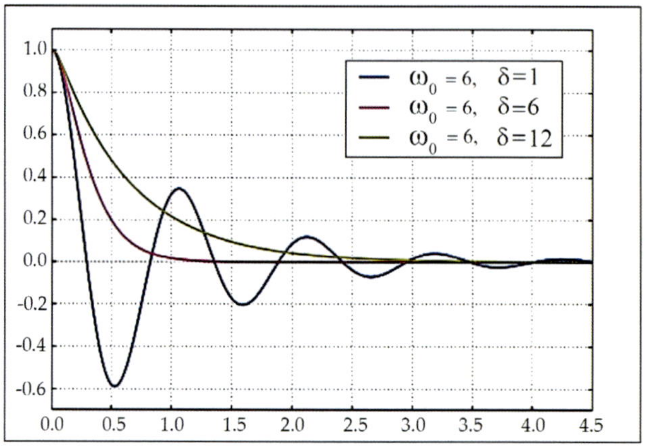

Figure 6.7. The relationship between the undamped and damped natural frequencies.

Source: Difference Between, Creative Commons License.

2. **Critical Damping:** When c = c_c, there is no oscillation, and the solution of Eqn. (6.14) is:

$$x = (A + Bt)e^{-ct/2m} \tag{6.18}$$

3. **Greater-Than-Critical Damping:** When $\zeta > 1$, the solution of Eqn. (6.14) is

$$x = e^{-ct/2m}\left(Ae^{\omega_n\sqrt{\zeta^2 - 1}\,t} + Be^{-\omega_n\sqrt{\zeta^2 - 1}\,t}\right) \tag{6.19}$$

If the system is moved from its equilibrium point, it prefers to return slowly because this motion is not oscillatory (Lancaster, 1960).

6.5. FORCED VIBRATION

When an outside force is applied to a single degree of freedom (SDOF) system, forced vibration occurs. The nature of this external factor may be non-periodic or periodic. In analyzing forced vibration in an SDOF system, it is necessary to ascertain its steady-state reaction or response once its frequency and amplitude have stabilized (Trifunac, 1972).

The forced vibration of an SDOF system has the following equation of motion:

$mx'' + cx' + k*x = F(t)$

Where c is the damping coefficient, m is the system mass, k is the spring constant, x is the system's displacement from its equilibrium point, and F(t) is the external factor that impacts the system.

Supposing a sinusoidal input force $F(t) = F_0\sin(\omega t)$, The motion equation may be expressed as:

$mx'' + cx' + kx = F_0\sin(\omega*t)$

The indeterminate coefficients approach can be used to find the answer to this equation. Considering a specific result of form $x(t) = X\sin(\omega t + \Phi)$, where Φ is the phase angle and x is the amplitude, this can be changed in the motion equation to get (Salawu & Williams, 1995):

$X = F_0/[(k-m\omega^2)^2 + (c\omega)^2]^{1/2}$

$\Phi = \arctan(c\omega/(k-m\omega^2))$

The system's amplitude X, phase angle (depending on the input force's frequency), spring constant (k), and damping coefficient (c). The frequency of an input force must match the system's inherent frequency for the amplitude X to be at its maximum, which is given by:

$\omega_n = (k/m)^{1/2}$

Resonance is the term used to describe this situation, which produces a strong amplitude response that can harm the system. Damping is provided to the framework to avoid resonance, which lowers the amplitude response at resonance (Yamakoshi et al., 1990).

The magnitude ratio is the relationship between the amplitude of a stable response and an applied force's amplitude.

$MR = X/F_0 = 1/[(k-m\omega^2)^2 + (c\omega)^2]^{\wedge 1/2}$

The formulae below provide the phase angle between the steady-state response and the input force.

$\varphi = \Phi - \omega*t$

These equations may be utilized to calculate an SDOF system's steady-state reaction to a sinusoidal input force. Fourier series research can locate the solution if such input force is not sinusoidal (Feldman, 1994) (Figure 6.8).

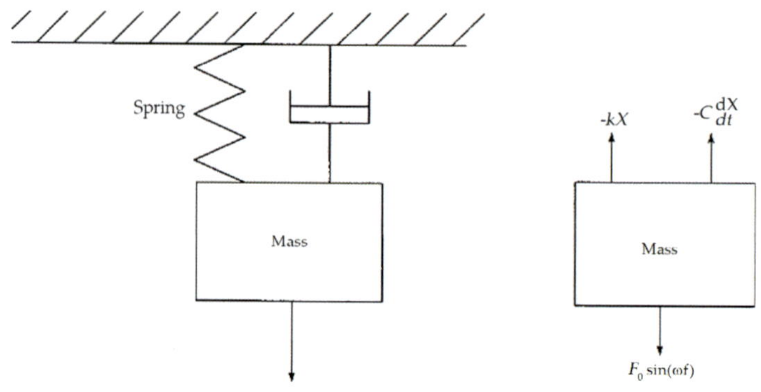

Figure 6.8. Illustration of forced vibration.

Source: Science Direct, Creative Commons License.

6.6. FORCED VIBRATION OF TWO DEGREES OF FREEDOM

"Two degrees of freedom systems" refers to a certain dynamic system that needs degrees of freedom or two independent coordinates to characterize its motion. The directions of the degrees of freedom can or cannot line up. A system with two degrees of freedom for the x and y directions is depicted in Figure 6.9(a). Another instance in which both degrees of freedom operate in a single direction but even on distinct masses is shown in Figure 6.9(b) (Yeh, 1966). There are two formulas of motion for a system with two degrees of freedom, each of which describes the motion of a single degree. These two equations are typically expressed as linked differential equations. These equations of motion may be applied to identify two natural frequencies, or phases, for such a system providing a harmonic solution for every coordinate (Cao & Sarkar, 2012).

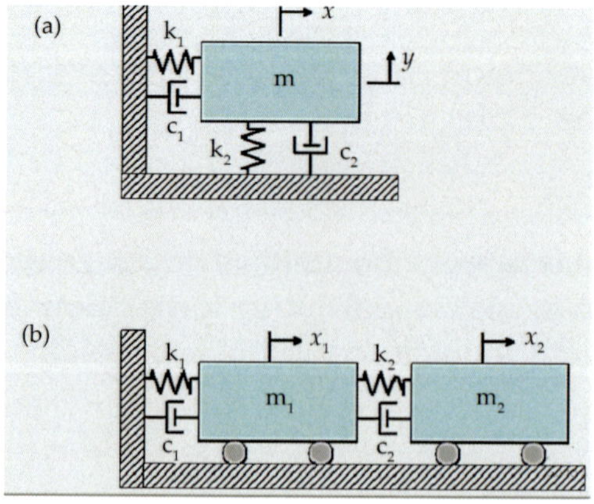

Figure 6.9. Schematic of system with two degrees of freedom for the x and y directions.

Source: Yeh, Creative Commons License.

1. **Obtaining the Equations of Motion:** Newton's second law may be applied to determine the motion equations for one system with two degrees of freedom. Think about the system in Figure 6.9(b). Measuring from the equilibrium positions of every mass, the coordinates $x_1(t)$ and $x_2(t)$ perfectly characterize the movement of this system. Upon masses of m_1 and m_2, respectively, external forces $F_1(t)$ or $F_2(t)$ is at work (Jauvtis & Williamson, 2003) (Figure 6.10).

We create free-body diagrams for every mass using Newton's second law, as seen in Figure 6.10.

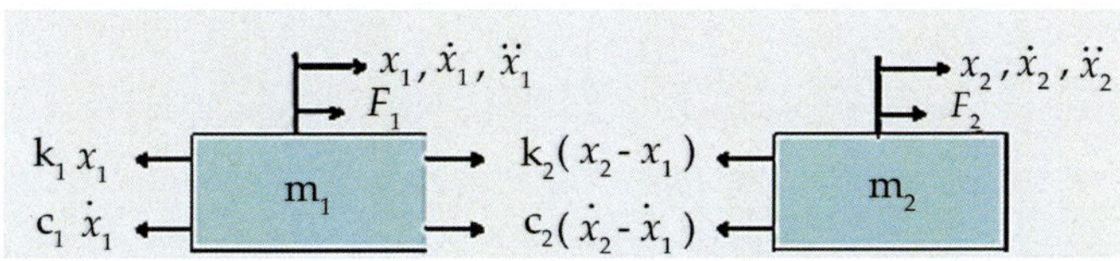

Figure 6.10. Picture of the two-degree-of-freedom systems with free-body diagrams for such masses.

Source: Zhao et al., Creative Commons License.

The following free-body diagrams make it simple to find the motion equations:

$$m_1 \cdot \ddot{x}_1 + (c_1 + c_2) \cdot \dot{x}_1 - c_2 \cdot \dot{x}_2 + (k_1 + k_2) \cdot x_1 - k_2 \cdot x_2 = F_1$$
$$m_2 \cdot \ddot{x}_2 - c_2 \cdot \dot{x}_1 + c_2 \cdot \dot{x}_2 - k_2 \cdot x_1 - k_2 \cdot x_2 = F_2$$

The differential equations remain coupled because both equations include the coordinates x_1 and x_2, as you may have seen. This physically implies that one mass's movement influences the other's movement. The movement equations can be usefully transformed into matrix form (Jeon & Gharib, 2001):

$$[m]\{\ddot{x}(t)\} + [c]\{\dot{x}(t)\} + [k]\{x(t)\} = \{f(t)\}$$

$$[m] = \begin{bmatrix} m_1 & 0 \\ 0 & m_2 \end{bmatrix}$$

Where, $[c] = \begin{bmatrix} c_1 + c_2 & -c_2 \\ -c_2 & c_2 \end{bmatrix}$

$$[k] = \begin{bmatrix} k_1 + k_2 & -k_2 \\ -k_2 & k_2 \end{bmatrix}$$

And, $\{x\} = \begin{Bmatrix} x_1 \\ x_2 \end{Bmatrix}$

$$\{F\} = \begin{Bmatrix} F_1 \\ F_2 \end{Bmatrix}$$

The squareness and dimension of damping, mass, and stiffness matrices are proportional to the number of freedom degrees (which, in this case, is 2). Often symmetrical matrices should also be noted.

The motion equations for only a general two-degree-of-freedom system may take the following form:

$$\begin{bmatrix} m_{11} & m_{12} \\ m_{12} & m_{22} \end{bmatrix} \begin{Bmatrix} \ddot{x}_1 \\ \ddot{x}_2 \end{Bmatrix} + \begin{bmatrix} c_{11} & c_{12} \\ c_{12} & c_{22} \end{bmatrix} \begin{Bmatrix} \ddot{x}_1 \\ \ddot{x}_2 \end{Bmatrix} + \begin{bmatrix} k_{11} & k_{12} \\ k_{12} & k_{22} \end{bmatrix} \begin{Bmatrix} \ddot{x}_1 \\ \ddot{x}_2 \end{Bmatrix} = \begin{Bmatrix} F_1 \\ F_2 \end{Bmatrix}$$

We represent the external forces as follows, assuming that they have a harmonic nature:

$$F_j(t) = F_{j0} \cdot e^{i \cdot \omega \cdot t}, \quad j = 1, 2$$

Hence the solutions at steady state as:

$$x_j(t) = x_{j0} \cdot e^{i \cdot \omega \cdot t}, \quad j = 1, 2$$

The ω dynamic properties of the system are represented by X_1 and X_2, which are complex variables that rely on. The following are the equations when expressed in terms of frequency:

$$\begin{bmatrix} \left(-\omega^2 m_{11} + i \cdot \omega \cdot c_{11} + k_{11}\right) & \left(-\omega^2 m_{12} + i \cdot \omega \cdot c_{12} + k_{12}\right) \\ \left(-\omega^2 m_{12} + i \cdot \omega \cdot c_{12} + k_{12}\right) & \left(-\omega^2 m_{22} + i \cdot \omega \cdot c_{22} + k_{22}\right) \end{bmatrix} \begin{Bmatrix} X_1 \\ X_2 \end{Bmatrix} = \begin{Bmatrix} F_{10} \\ F_{20} \end{Bmatrix}$$

Let mechanical impedance be:

$$Z_{rs}(\omega) = -\omega^2 \cdot m_{rs} + i \cdot \omega \cdot c_{rs} + k_{rs}, \quad r, s = 1, 2$$

The impedance matrix is defined as:

$$[Z(\omega)] = \begin{bmatrix} Z_{11}(\omega) & Z_{12}(\omega) \\ Z_{12}(\omega) & Z_{22}(\omega) \end{bmatrix}$$

And the system equations can be written as (Dahl et al., 2006).

$$[Z(\omega)]\{X(\omega)\} = \{F_0(\omega)\} \rightarrow \{X(\omega)\} = [Z(\omega)]^{-1}\{F_0(\omega)\} = [H(\omega)]\{F_0(\omega)\}$$

$$\{X(\omega)\} = [Z(\omega)]^{-1}\{F(\omega)\} = [H(\omega)]\{F(\omega)\}$$
$$[H(\omega)] = [Z(\omega)]^{-1}$$

The frequency response matrix is the inverse of the impedance matrix. $[H(\omega)]$.

The solution for every response is:

$$x_1(\omega) = H_{11}(\omega)F_1(\omega) + H_{12}(\omega)F_2(\omega)$$
$$x_2(\omega) = H_{21}(\omega)F_1(\omega) + H_{22}(\omega)F_2(\omega)$$

$H_{ij}(\omega)$is a complex exponential function of frequency and shows the connection between a force that acts at degree j and a response at a degree I of freedom (Haddow et al., 1984).

Frequency response function in the usual sense H_{ij} (ω) for a 2 This is a magnitude/phase representation of the DOF system (Gsell et al., 2016).

TIPS

Understanding the dynamics of the system and how the forces interact is key to predicting and controlling the resulting motion.

6.7. FREE VIBRATION OF MULTI-DEGREE OF FREEDOM (DOF) SYSTEMS

6.7.1. System Without Damping

The two-DOF system's equation of motion in free vibration without an external force is.

mu +ku = 0

The solutions, including an initial condition, are just mass displacements.

$$\mathbf{u} = \mathbf{u}(0) \text{ and } \dot{\mathbf{u}} = \dot{\mathbf{u}}(0)$$

Every mass will move as illustrated if any two-DOF system is allowed to vibrate with any initial displacement. These are not simple harmonics since the free SDF system vibrations and deflection form change over time as the ratio of u_1/u_2 changes (Mead, 1975) (Figure 6.11).

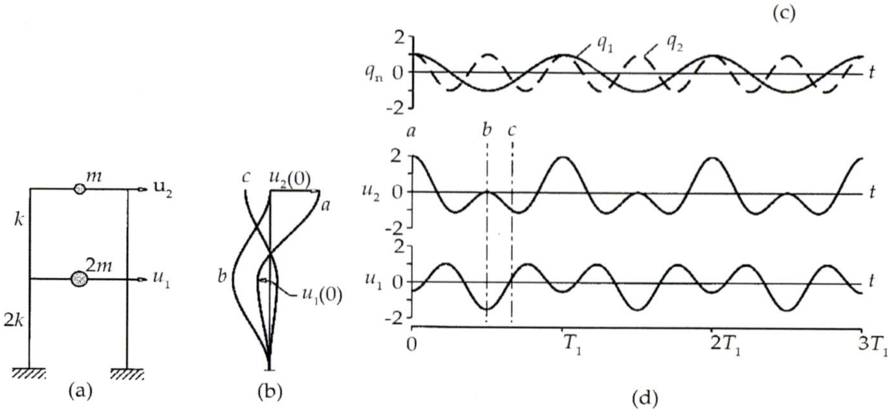

Figure 6.11. Diagram of free vibration of an undamped system due to arbitrary initial displacement: (a) two-story frame; (b) deflected shapes at time instants a, b, and c; (c) modal coordinated $q_n(t)$; and (d) displacement history.

If an undamped structure starts free vibration properly, displacement shapes will experience simple harmonic motion without changing their deflected shape. Each floor would simultaneously reach maximum displacement while maintaining its basic shape (Olmos & Roesset, 2010).

Both floors of the first form above move in the same direction. These two floors of a second form move in opposition to one another. "Node" is the term for the position when displacement stays zero. "Natural modes of vibration" are the names given to these forms. There are more modes when there are more degrees of freedom (Coron, 1998).

Every mode causes all floors to vibrate at a single frequency, known as the "natural circular frequency of vibration." (ω_n) or "natural cyclic frequency of vibration" (f_n), and the term "natural period of vibration" refers to the duration of one cycle.

$$T_n = \frac{2\pi}{\omega_n} \qquad f_n = \frac{1}{T_n}$$

The fundamental natural frequency, symbolized by the number 1, is the lowest. The fundamental natural period (T_1) of vibration, which is the lengthiest natural period, is what it refers to (Choi et al., 2007).

6.7.2. Natural Vibration Frequencies and Modes

Displacement vector **u**(t) can be expressed as a product of the scalar quantity modal coordinates and the mode shape vector n.

$$\mathbf{u}(t) = q_n(t)\phi_n$$

The modal analysis vector does not change over time, but the modal coordinate changes. The modal coordinate fluctuates like a straightforward harmonic having frequency equal to a natural frequency during free vibration (Van Rensburg et al., 2006).

$$q_n(t) = A_n \cos\omega_n t + B_n \sin\omega_n t$$

Therefore,

$$\mathbf{u}(t) = \phi_n\left[A_n \cos\omega_n t + B_n \sin\omega_n t\right]$$

When we plug this into the equation of motion, we get.

$$\left[-\omega_n^2 \mathbf{m}\phi_n + \mathbf{k}\phi_n\right]q_n(t) = \mathbf{0}$$

There are two ways to solve this equation:

1. $q_n(t) = 0$ Yet, because this is not very beneficial, it is referred to as a trivial solution.
2. $-\omega_n^2 \mathbf{m}\phi_n + \mathbf{k}\phi_n = \mathbf{0}$ or $\mathbf{k}\phi_n = \omega_n^2 \mathbf{m}\phi_n$

This is referred to as a matrix eigenvalue problem. We understand the matrices k and m and want to find ω_n^2 and φ_n (Nayfeh, 1995).

$$\mathbf{k}\phi_n - \omega_n^2 \mathbf{m}\phi_n = \mathbf{0} \qquad \left[\mathbf{k} - \omega_n^2 \mathbf{m}\right]\phi_n = \mathbf{0}$$

This equation, yet again, could be satisfied by either $\varphi_n = \mathbf{0}$ (trivial) or

$$\det\left[\mathbf{k} - \omega_n^2 \mathbf{m}\right] = 0$$

In terms of, this determinant is just a polynomial of order N. ω_n^2. This is referred to as a "characteristic equation or frequency equation." "It has N real and positive roots for ω_n^2 k and mare are symmetric as well as positive definite matrices.

The N roots for ω_n^2 are known as eigenvalues, or These have distinctive values and correspond to a natural frequency of N vibration modes (Nayfeh et al., 1995).

For each value of ω_n^2, φ_n the equation can be used to solve the problem.

$$\left[\mathbf{k} - \omega_n^2 \mathbf{m}\right]\phi_n = \mathbf{0},$$

They are not unique, however, because multiple vectors formed by scaling a φn solution would also fulfill the equation. The shape of this vector φ_n is called a natural mode of vibration or mode shape. It is also known as the eigenvector, corresponding to the eigenvalue ω_n^2 (Chen et al., 2008).

Did you Know?

Each natural frequency corresponds to a unique mode shape, which describes the pattern of vibration exhibited by the structure.

6.7.3. Mode and Spectral Matrices

Several mode shapes could be combined in a matrix to generate a compact form.

$$\mathbf{\Phi} = \left[\phi_{jn}\right] = \begin{bmatrix} \phi_{11} & \phi_{12} & \cdots & \phi_{1N} \\ \phi_{21} & \phi_{22} & \cdots & \phi_{2N} \\ \vdots & \vdots & \ddots & \vdots \\ \phi_{N1} & \phi_{N2} & \cdots & \phi_{NN} \end{bmatrix}$$

The matrix Φ is called the modal matrix for an eigenvalue problem. The N eigenvalues ω_n^2 may be assembled into a diagonal matrix Ω^2, the eigenvalue problem's spectral matrix (Held, 1985).

$$\mathbf{\Omega}^2 = \begin{bmatrix} \omega_1^2 & & & \\ & \omega_2^2 & & \\ & & \ddots & \\ & & & \omega_N^2 \end{bmatrix}$$

In vector form, the eigenvalue issue

$$\mathbf{k}\phi_n = \omega_n^2 \mathbf{m}\phi_n$$

They can be written in a matrix format.

$$\mathbf{k\Phi} = \mathbf{m\Phi\Omega}^2$$

6.7.4. Orthogonality of Modes

The orthogonality criteria for natural phases corresponding to distinct natural frequencies could (Fraser et al., 1976).

when; $\omega_n \neq \omega_r$,

$$\phi_n^T \mathbf{k} \phi_r = 0 \qquad \phi_n^T \mathbf{m} \phi_r = 0$$

It can be demonstrated as follows: From (Mestas et al., 2000).

$$\mathbf{k}\phi_n = \omega_n^2 \mathbf{m} \phi_n$$

Premultiply by φ_r^T

$$\phi_r^T \mathbf{k} \phi_n = \omega_n^2 \phi_r^T \mathbf{m} \phi_n$$

Because the above is scalar, we could transpose the right side (Fraser, 1976).

$$\phi_n^T \mathbf{k} \phi_r = \omega_n^2 \phi_n^T \mathbf{m} \phi_r$$

Let us begin with the rth mode,

$$\mathbf{k}\phi_r = \omega_r^2 \mathbf{m} \phi_r$$

Premultiply by φ_n^T, we get:

$$\phi_n^T \mathbf{k} \phi_r = \omega_r^2 \phi_n^T \mathbf{m} \phi_r$$

Subtract the two equations, and we get:

$$\left(\omega_n^2 - \omega_r^2\right) \phi_n^T \mathbf{m} \phi_r = 0$$

When $\omega_n^2 \neq \omega_r^2$, $\phi_n^T \mathbf{m} \phi_r = 0$ and also $\phi_n^T \mathbf{k} \phi_r = 0$.

The orthogonality condition indicates that the matrices are orthogonal.

$$\mathbf{K} \equiv \mathbf{\Phi}^T \mathbf{k} \mathbf{\Phi} \qquad \mathbf{M} \equiv \mathbf{\Phi}^T \mathbf{m} \mathbf{\Phi}$$

Are diagonal where the diagonal elements are

$$K_n = \phi_n^T \mathbf{k} \phi_n \qquad M_n = \phi_n^T \mathbf{m} \phi_n$$

Because k and m are positive definite, so K_n and M_n are positive and are related by (Gilar et al., 2005)

$$K_n = \omega_n^2 M_n$$

by

$$K_n = \phi_n^T \mathbf{k} \phi_n = \phi_n^T \left(\omega_n^2 \mathbf{m} \phi_n \right) = \omega_n^2 \left(\phi_n^T \mathbf{m} \phi_n \right) = \omega_n^2 M_n$$

6.8. VIBRATION CONTROL

To effectively combat the harmful consequences of vibrations, we must first understand the causes of their emergence and growth. It calls for thoroughly examining the operational dynamics of the machines and devices under consideration. Due to the intricate architecture of the machinery and devices, such tasks are exceedingly challenging. Machine units are complex systems comprising numerous pieces, multiple degrees of freedom, and branching closed or opened kinematics. These are distinguished by elastic components, elements having variable mass and clearances in kinematic pairs, nonholonomic or holonomic limitations, etc. (Alkhatib & Golnaraghi, 2003).

Machines and gadgets can generate mechanical vibrations from various sources, but locating and identifying these sources can be exceedingly challenging. Many factors contribute to vibrations that interfere with the regular operation of machinery. A few of them are the results of technical realization—the other results from subpar elemental work, improper assembly, component wear, etc. In addition, a few causes result from external environmental influences (Franchek et al., 1996).

The aforementioned myriad vibration sources that occur while machines operate impact their components. This impact is consistent with regular, nearly regular, or unpredictable inputs. The above inputs interact with other agents, such as a static load, a compounded stress state, and a region with changeable temperature. Resonant states, dynamically critical states, including states with unstable vibration motions, can all develop under these circumstances (Thenozhi & Yu, 2013). Machine components frequently develop strain cracks due to high, changing strains accompanying vibration phenomena, ultimately leading to their breakdown. The vibrations affect the robustness of machines, disrupt the programmed movement of machine parts, provide a noise source, and increase friction. These also serve as the reason for high mechanical energy loss, which reduces machine and device performance, efficiency, and durability simultaneously. Furthermore, the vibrations hurt the environment and the machinery's workers (Jalili, 2002).

Considering the factors above, it is possible to say that the most significant hazardous occurrences that follow machine vibrations are:

- Disruptions to the regular operation of machinery (Zhang et al., 2017).
- A deterioration in the durability of machines and gadgets, an adverse impact of vibrations upon that environment,
- Noise emission (Brennan, 1997).

It indicates that the following are the primary goals of the machine vibration control:

- The best protection for machine operations;
- Minimizing damaging impacts, such as vibrations in the surroundings 2017).

The mathematical modeling stage is the first phase of assessing and synthesizing dynamic processes inside the device. The FEM is among the most widely used modeling techniques. By employing this technique, a substantial number of tiny elements are used to replace the device structure. The stiffness (Ke), load (pe), inertia (Me), and damping (Be) matrices are then produced for every element using the coordinates that were previously selected. Moreover, absolute coordinates are selected, and their complete matrices K, M, P, and B are calculated. The system's mathematical model has the following shape (He & Ge, 2015).

$$Mx + Bx + Kx = P$$

where; x is the vector of absolute coordinates (Kowal, 1996).

Little damping values resulting in matrix B's components taking the same values are frequently characteristic of mechanical plants. Small input amplitudes result in a system's resonance vibrations if input P's frequency equals one of the inherent frequencies of a plant. The following issues must be considered to limit vibrations and avoid unwanted disturbances (Daley et al., 2004).

- Alteration of the P vector's coordinates (i.e., to eliminate or decrease values of input forces),
- selection of the proper matrix elements M, K, B (i.e., to increase natural vibrations frequency) (Stanway, 1996).

6.9. ACTIVE VERSUS PASSIVE CONTROL

If a precise structure is exposed to variable thermal conditions, it will distort because of the heat gradients unless it is precisely engineered. Building the structure out of a high thermal stability composite material is one technique to avoid this; it is a passive method. One alternative is using an active structure with several sensors and actuators and a feedback loop. In this instance, we take advantage of feedback's primary benefit: to lessen the output's sensitivity to parameter changes and the impact of disturbances within the controlling system's bandwidth. Active structures might be more affordable or lighter than passive structures, which operate similarly depending on the situation. Like in the following example, they might have capabilities that no passive structure could (Chrastil et al., 2012). The Hubble space telescope was created because, up until a few years ago, it was widely believed that environmental turbulence would significantly restrict the resolution of earth-based observatories.

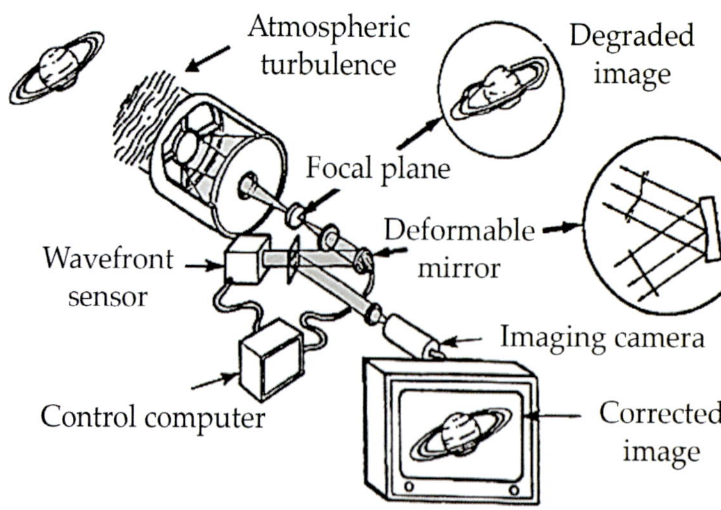

Figure 6.12. Illustration of principle of adaptive optics for compensating atmospheric turbulence.

Source: G. Rousset-ONERA, Creative Commons License.

The disturbances resulting from atmospheric turbulence upon that optical wavefront originating from astronomical objects can now be corrected in real-time, allowing us to increase the telescope's maximum resolution by such an order of magnitude to the point of limitation imposed by diffraction (Deak et al., 2012). A deformable mirror attached to a group of actuators corrects (Figure 6.12). The control computer provides the form of a deformable mirror necessary to repair this error once the wavefront sensor identifies the phase difference within the turbulent wavefront. During ground-based astronomy, adaptive optics is now considered a standard component (Benware, 1984).

The primary mirror of huge telescopes with a radius of 8 m or more is another example of such an active structure that has proven advantageous to astronomy. Massive primary mirrors are exceedingly challenging to produce and put together. Maintain its proper shape despite variable gravity loads experienced during star monitoring and the dynamic stresses from the wind; a passive mirror should be thermostable and extremely stiff. There must be two active options to it (The first method, used on the very large telescope (VLT) at ESO in Paranal, Chile, entails having a primary mirror that is reasonably flexible and is attached at the back to a group of about a 100 actuators (Krishna et al., 2008). Like the previous illustration, the control method evaluates the optical modes' perturbation amplitude using an image analyzer before computing a correction to lessen its impact and applying it through the actuators. With both the image analyzer, it is possible to experimentally ascertain the effect matrix J between actuator forces f and the optical modal amplitudes w of a wave frontal errors (Rachels, 1975).

$$w = Jf \quad (6.20)$$

J is a rectangular matrix since there are more actuators than interesting optical modes. The corrective forces could be derived from the evaluated phase errors w.

$$f^{*} = J^{T}(JJ^{T})^{-1}w^{*} \quad (6.21)$$

$J^{T}(JJ^{T})^{-1}$ is the pseudo-inverse of the rectangular matrix J. This is the least possible norm solution to Equ. (6.20). The second option involves utilizing a segmented primary mirror, which the Keck observatory in Mauna Kea, Hawaii, has done. Such a design may have benefits, including lighter weight, reduced cost, and simplicity in production and assembly.

The control system obtains the visual properties of a monolithic mirror (by cophasing the segments), accounts for gravity and wind disturbances, and minimizes the effect of the telescope dynamics upon that optical performance. Every segment does have a hexagonal structure. Also, it is equipped with three computer-controlled freedom degrees (tilt and piston) and six edge sensors trying to measure the relative displacements concerning the two neighboring segments. A third instance, again in astronomy, is the upcoming interferometric missions. The goal is to produce a resolution that can only be attained through a significantly larger monolithic telescope by using a selection of smaller telescopes like an interferometer. A potential spacecraft design for a similar interferometric mission is shown in Figure 6.13; it comprises a main truss holding a group of telescopes that can aim in different directions. The optical routes between the various telescopes and the beam combiner are precisely regulated by optical delay lines using information from a wavefront sensor. The relative locations of a telescope are maintained by such sophisticated metrology. Normally, there could be up to 50 meters between telescopes, depending on the sequence (Desmond, 2000) (Figure 6.13).

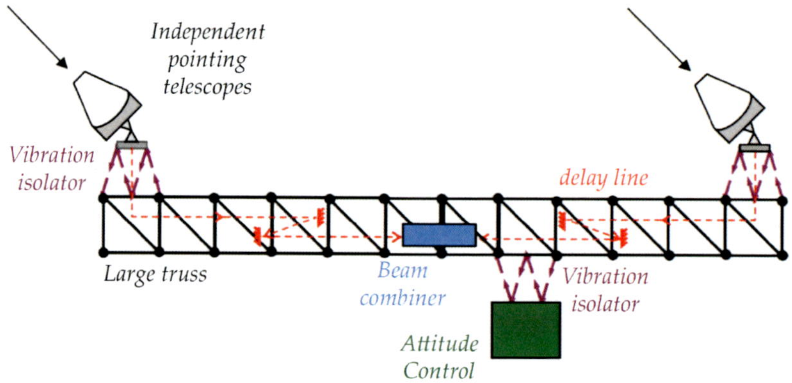

Figure 6.13. Picture of future interferometric mission schematic view.

Source: A. Preumont, Creative Commons License.

The individual telescopes' pointing errors are as small as a few nano radians. The maximum error allowed mostly on the length of the optical path is just a few nanometers (i.e., one order of magnitude better than the Hubble space telescope). A precise monolithic construction cannot meet such exacting geometrical demands within a harsh space environment; active methods, as proposed in Figure 6.13, must be used. The key requirement for the supporting truss is stability rather than precision, with the wide-band vibration exclusion control scheme of each telescope and the optical delay lines taking care of the accuracy of an optical route (described below). In deployable constructions, geometric stability also involves thermal stability, vibration dampening, and pre-stressing the gaps (a critical issue for deployable trusses). In addition to these geometric constraints, this spacecraft will be launched in deep space instead of in low orbit to achieve maximum sensitivity; that weight issue is, therefore, very crucial (Mohan et al., 2011).

The six d.o.f. A vibration isolator only at the interface between the attitude control module and the supporting truss is another intriguing subsystem required to meet the strict specifications. This isolator needs to allow the lower frequencies of attitude control torque to be transferred while filtering out its high-frequency disturbances caused by unbalanced centrifugal forces within reaction wheels. When the truss meets the individual telescopes, an additional vibration isolator can be employed, maybe in conjunction with telescopic steering (Rogelberg et al., 2003). The optical delay line, the third component important to active control, comprises a highly accurate single-degree-of-freedom translational method supporting a mirror. Its purpose is to control the length of the optical system between each telescope and the beam combiner to ensure that these ranges are maintained constant to a fraction of a wavelength (e.g., λ/20). These examples largely focused on performance. Yet, it is anticipated to be a steadily increasing number of applications wherein active solutions will grow to be less expensive than passive ones while maintaining the same performance level, while technology advances as low-cost electronic components become more readily available, e.g., readers shouldn't draw the wrong conclusions, like the fact that active remains are preferable and that a good control system could make up for poor design (Chaudet et al., 2005). A badly designed will typically remain terrible whether it is active or not because an active solution can be considered once all the passive options have been tried. Always remember that feedback control could make up for external forces only within a certain frequency range, referred to as the control system's bandwidth. Always forget that such a control mechanism amplifies the disturbance outside of the bandwidth (Alexiadou & Doron, 2012).

6.10. VIBRATION SUPPRESSION

Regarding amplitude, mechanical vibrations range between meters (civil engineering) to nanometer range (precision engineering). They may negatively affect systems (Onoda et al., 1991).

6.10.1. Failure

Vibration-induced structural collapse can be caused by fatigue, stability owing to operational circumstances (such as bridge flutter during wind stimulation), or excessive strain following transient events (such as a building's response to earthquakes) (mechanical parts in machines) (Pinto & Mantel, 1990).

6.10.2. Comfort

Many situations exist where vibrations are uncomfortable, including helicopter vibrations, noise, vehicle suspensions, and wind-induced building sway (Kolcaba & Kolcaba, 1991).

6.10.3. Operation of Precision Devices

Precision engineering uses various methods to limit mechanical vibrations, particularly optical systems, severely. Examples include precision machinery, wafer steppers, and telescopes. Micro vibrations impacting the various components of the optical system hinder the performance of big interferometers like the VLTI. Due to different disturbance sources like the temperature gradient,

it won't be easy to construct lightweight segmental telescopes (either in space or on the ground) in their final configuration with only an accuracy of a fraction of both a wavelength (which dominates the space environment). These systems won't exist if the reflector form is not actively controlled (Aguado et al., 2018).

Depending on the issue, there are several strategies to reduce vibration; the most popular ones include damping, stiffening, and isolation. Shifting the resonant frequencies of the structure outside the excitation region constitutes stiffening.

Damping is the process of lowering resonance peaks by releasing vibrational energy. The goal of isolation is to stop disturbances from spreading to vulnerable areas of the systems (Xue et al., 2021).

They use kinetic energy to drive dynamic vibration absorbers or passively dampen with hydraulic dampers, elastomers, eddy currents, or hysteretic components (DVA) (Gao et al., 2020). Transducers may also convert vibrational energy into electrical power, which may be saved or lost in electrical networks (energy harvesting). Semi-active devices, also known as semi-passive devices, are now readily accessible; they are composed of passive components with programmable characteristics (Kazdaridis et al., 2020).

Famous examples include piezoelectric transducers with switching electrical networks and the magnetorheological (MR) fluid damper. Because of their non-linear behavior, semi-active devices may transfer energy from one frequency to another (Hung et al., 2022). Still, they are essentially passive and, unlike active devices, cannot cause the system to become unstable. Semi-active devices are thus less susceptible to power outages. Active control, which entails a collection of sensors (strain, velocity, acceleration, and force), a collection of actuators (force, inertial, and strain), as well as a control algorithm, could be utilized when high performance is required (feedback or feedforward) (Gao et al., 2020). Among the key topics of such a book is active damping. The configuration of the actuators and sensors, how to ensure reliability and durability (e.g., collocated actuator/sensor pairs), and the energy requirements will frequently influence the dimensions of the actuators. The project's cost is just one issue that arises when creating an active control system (Hung et al., 2021) (Figure 6.14).

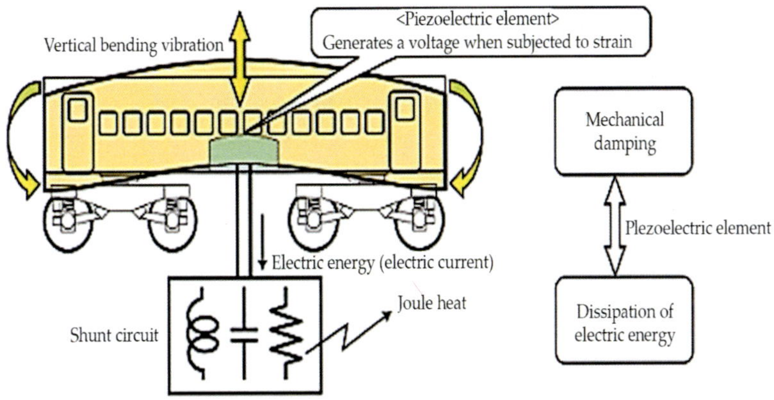

Figure 6.14. Illustration of vibration suppression.

Source: Madalina Dumitriu, Creative Commons License.

6.11. SMART MATERIALS AND STRUCTURES

An active structure has several sensors and actuators connected by a controller; if the bandwidth of both controllers covers any of the structure's vibration modes, its dynamic behavior should be considered. The collection of sensors and actuators could be individually positioned on the structure. The unique characteristic of smart structures is that actuators and sensors are frequently dispersed and highly integrated within the structure, making it hard to model them separately (Mackerle, 2003). Moreover, in some applications, such as vibroacoustic, the behavior of the structure is tightly connected to the environment, necessitating coupled modeling as well. Mechanically speaking, the elastic constants that link stress and strain and the coefficient of thermal expansion that links strain to temperature completely characterize the properties of traditional structural materials. Materials may also experience strain due to various causes, including temperature, an applied electric, or even a magnetic field (Sun et al., 2016).

The following is a quick summary of the most well-known smart materials:

1. Shape memory alloys (SMA) enable recovery of up to 5% strain following temperature-induced phase shift. SMA is most suited to one-way operations like deployment, even though two methods are conceivable after education. In any event, due to the challenges of cooling, they may only be employed for low-accuracy and low-frequency applications. Another issue with thermal cycling is fatigue. NITINOL is the most popular SMA, so this book would not mention

SMA because it is rarely used during vibration control applications (Sun et al., 2016).

2. Piezoelectric materials could be employed as actuators and sensors because of their 0.1% recovery strain under such an electric field. Polymers and Ceramics are the two main categories of piezoelectric materials utilized in vibration control. Due to their excessive voltage requirements and restricted control capabilities, the piezopolymers are primarily utilized as sensors, the most well-known being polyvinylidene fluoride (PV DF or PV F2). Piezoceramics are widely utilized as sensors and actuators, especially for ultrasonic applications covering a variety of frequency ranges. They are ideally suited for high accuracy in the nanoscale range (1 nm = 109 m). Lead zirconate titanate (PZT) is the most well-known piezoceramic; PZT patches could be bonded or co-fired upon that supporting framework (Sanders et al., 2004).
3. Magneto-strictive materials get a recoverable strain of 0.15% in a magnetic field; the material responds at its best under these conditions and is loaded in a compressive manner.
4. Magneto strictive actuators have a long-life span and may be utilized as load-bearing components (in compression only). Moreover, they could be utilized in highly precise applications. The most popular is TERFENOL-D, which can substitute for PZT in some situations (sonar) (Spillman et al., 1996).
5. MR fluids comprise viscous fluids with magnetic substances that take the shape of micron-sized particles. Whenever a magnetic field is applied to a fluid, the particles form columnar formations requiring a modest shear force to start flowing. Its impact is swift and reversible (response time of the order of milliseconds). Certain fluids, known as electro-rheological (ER), behave similarly inside an electric field. However, their performances are presently inferior to MR fluids (limited by electromagnetic field breakdown). Semi-active devices employ MR and ER fluids (Mackerle, 1998).

Just a sample of the forthcoming smart materials is shown in this brief list: In a few years, phase-changing materials, which are now under development and deliver an order of magnitude greater recoverable strain than piezoceramics (1%) underneath an electrical or magnetic field, are anticipated to be commercially accessible. For applications requiring great strain and low stiffness, electro-active polymeric also is steadily becoming more prevalent (Vinyas, 2021).

Many instruments can measure location, speed, acceleration, and strain, but more are coming, especially in opt mechanical fields. Inductive, al methods capacitive (laser interferometer) may be used to detect displacements; the last two have such a resolution inside the nanoscale range. While being relatively common, piezoelectric accelerometers could measure a d.c. Element.

Piezoceramics, piezopolymers, strain gages, and fiber optics are all methods for measuring strain. They have a huge potential for healthcare monitoring and may be integrated into a structure to provide a global average estimate of a deformation. Piezopolymers can be configured to respond to several modal analyzes (modal filters) (Forsythe & Johnson, 2017).

6.12. CONTROL STRATEGIES

Feedback and feedforward are two strategies for disturbance rejection that are drastically different. Although this book focuses only on feedback control, it is vital to highlight the key aspects of both strategies so that the user may choose which is best for a particular application (Crago et al., 1996).

6.12.1. Feedback

The feedback concept is illustrated in Figure 6.15; the system's output y is compared with the reference input r. System G utilizes the error signal, e = ry, through a compensator H(s) (s). The design challenge is finding the right compensator H(s) to ensure that the closed-loop system remains stable and operates as intended (Hattie & Timperley, 2007).

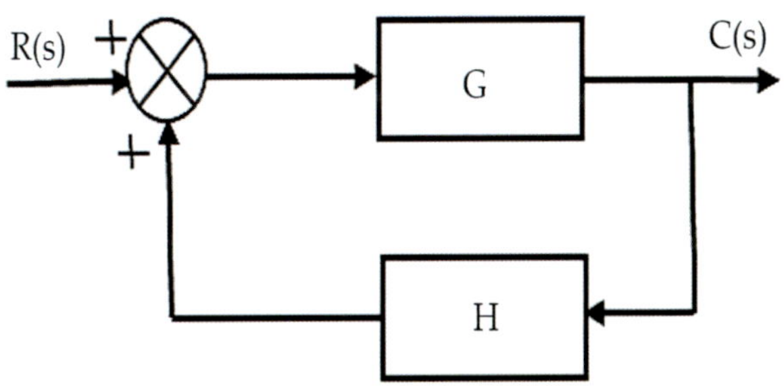

Figure 6.15. Image of principle of feedback control.

Source: Tutorials Point, Creative Commons License.

Active damping and model-based feedback are two different but relatively complimentary uses of feedback control inside the control of weakly damped systems (Ilgen et al., 1979).

Active damping aims to lessen the influence of resonating peaks upon that structure's response. From:

$$\frac{y(s)}{d(s)} = \frac{1}{1+GH} \quad (6.22)$$

It calls for GH 1 close to the resonances. With perfect dynamics and a collocated actuator with a sensor, active damping is frequently performed with modest improvements and with the added benefit of assured stability without needing a structural model. Naturally, the dynamics between sensors and actuators are always limited, and the bandwidth of every active damping system is finite.

The control aims might be more ambitious because we may want to maintain a control variable (such as an antenna's location or pointing) at a specific value r despite outside disturbances d in a certain frequency range. We are using the prior formula (Greller & Herold, 1975).

$$F(s) = \frac{y(s)}{r(s)} = \frac{GH}{1+GH} \quad (6.23)$$

We can immediately see that this necessitates high quantities of GH in the desired frequency range. If GH 1, the closed-loop transfer must occur.

When function F(s) is almost 1, the output y closely follows the input r. It also provides disturbance rejection inside the control system's bandwidth, according to Eqn. (6.23). In general, we require a more complex approach to accomplish this, comprising a mathematical system model that, at most, can serve as a low-dimensional approximation of the real system G(s). Several methods for locating the proper compensator will be discussed in this work, but only the most straightforward and well-established ones. They all share the following characteristics (Harvey, 2003).

1. The model's precision restricts the controlling system's bandwidth ωc.; The flexible forms outside are occasionally destabilized ω_c (residual modes). Spillover refers to the phenomenon wherein the net damping of residual modes diminishes as the bandwidth rises.
2. The rejecting of disturbances inside the controlling system's bandwidth is always made up for by amplifying disturbances outside of the bandwidth (Adcroft, 2011).
3. To reliably retain the behavior of the continuous system if represented digitally, the sampling frequency ωs must always be two orders of magnitude higher than ωc. The overall bandwidth of a control scheme is thus subject to various hardware constraints (Harvey, 2003).

6.12.2. Feedforward

If a signal associated with the disturbance is present, feedforward adaptive filtering, which was first created for noise reduction, is an attractive approach to feedback for transient response (Kluger & Nir, 2010).

Yet, it is also quite effective in controlling vibrations (Fuller et al.). Figure 6.16 explains the basic idea. The technique depends on the presence of a reference signal linked to the primary disruption; this signal is then run through an adaptive filter, and its output is then utilized by the system using secondary sources. Reduce the error signal placed at a single or more crucial place; the filter coefficients are adjusted. The goal is to create a secondary disturbance that cancels out the initial disturbance's impact where the error sensor is located (Lamme & Roelfsema, 2000). The approach may thus be thought of as a localized one, as opposed to global feedback. In addition, there is no assurance that now the global reaction is likewise lowered at other sites. Unless such a single-mode fiber dominates the reaction, there must be areas in which the reaction could be magnified. In contrast to active damping, which can only reduce disturbances close to resonances, feedforward operates anyone at frequency. It aims to eliminate the disturbance by producing a second signal with the opposite order (Eldan & Shamir, 2016).

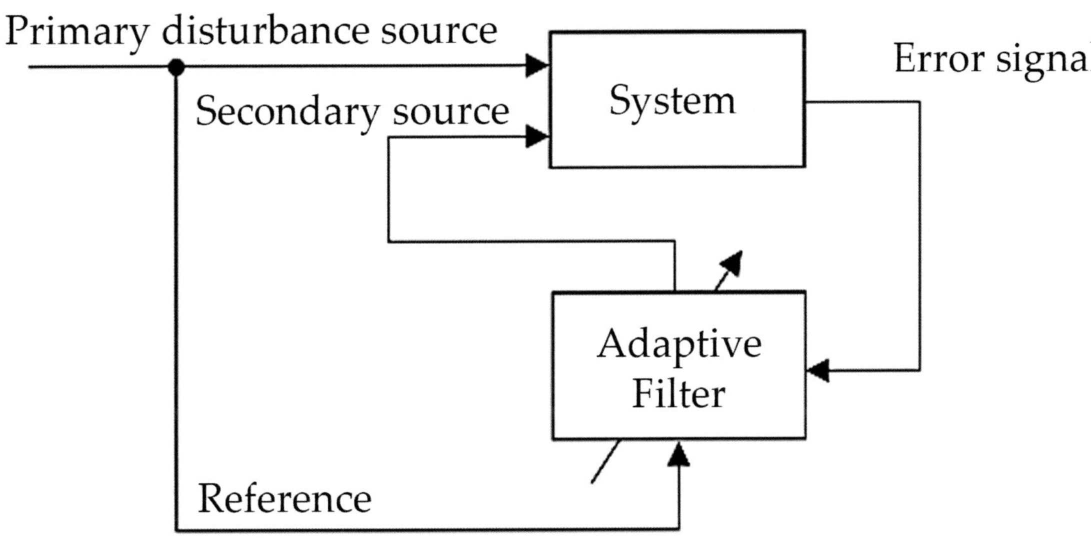

Figure 6.16. Illustration of principle of feedforward control.

Source: Optics Arizona, Creative Commons License.

Although the adaptation depends on the observed impulse response, the approach does not require a system model. Wide-band applications have additionally been reported, although narrow-band disruptions are where the method performs best. Feedforward control is more tolerant of phase lag than feedback and can operate at higher frequencies (a good rule of thumb is ω_c ‘ ω_s/10); This explains why it has had such great success in acoustics.

The absence of a reference signal associated with the disturbances is the primary drawback of feedforward adaptive filtering. Thus, a signal may easily be obtained from such a sensor placed on the perturbation's propagating route in various applications. Any impulse train produced by spinning machinery is responsible for machine-induced disruptions (Glorot & Bengio, 2010).

6.13. NON-LINEAR VIBRATION

Whenever a system's reaction is not inversely correlated with its stimulation's magnitude, a non-linear vibration occurs. Non-linear systems display complicated and frequently unpredictable behavior in contrast to linear systems, which respond to an external factor or disturbance in a known and well-defined manner. Several physical systems, particularly mechanical and electric, including biological ones, exhibit non-linear vibrations (Lamme & Roelfsema, 2000).

The existence of several resonant frequencies that can combine to create complex vibrational patterns constitutes one of the characteristics that distinguishes non-linear vibrations. The resonant frequency in linear systems is a constant value influenced by the system's physical characteristics, including its stiffness and mass.

However, the resonant frequency may vary in non-linear systems according to the excitation's magnitude and frequency (Seidler et al., 2004). This results in phenomena like subharmonic resonance, in which the reaction frequency is a small portion of the resonance frequencies, with frequency mixing, where various frequencies combine to generate a new frequency (Hornik et al., 1989).

Bifurcations, which are abrupt changes inside the system's behavior as parameter changes, are another feature of non-linear vibration. Bifurcations can cause the system to behave dramatically differently by causing new vibrational modes to develop or existing ones to vanish.

For instance, a non-linear system may move from one vibration mode to another when the frequency and amplitude of such an excitation are slightly altered (Eldan & Shamir, 2016).

Several real-world applications, like constructions subjected to significant deformation, electronic systems containing non-linear elements, and biological systems like the human heart, all contain non-linear vibrations. The developers and managers of non-linear systems must comprehend and anticipate their behavior. Numerical simulations are one method for investigating non-linear vibration as they can simulate the complex behavior of such systems and forecast how they will react to various inputs. Another strategy is experimental testing, which involves seeing and analyzing how a physical system behaves under various circumstances (Blum & Li, 1991).

Non-linear vibration also has practical applications, such as in the design of energy harvesting systems. Energy harvesting systems produce electrical energy by harnessing ambient vibrations, like those caused by equipment vibrations and wave motion. Non-linear systems are a desirable alternative for applications that harvest energy because they have a greater energy transfer rate and can generate more power than linear systems (Glorot & Bengio, 2010).

In summary, non-linear vibration is a sophisticated and intriguing phenomenon that occurs in various physical systems.

Although it might be challenging to forecast its behavior, it is vital to the development and functioning of real-world applications to comprehend and govern it. Further study and growth in this field may result in fresh perspectives and advances in structural design, bioengineering, and energy harvesting (Hornik, 1991).

6.14. APPLICATIONS OF VIBRATION

Vibration has several uses in various industries, including engineering, medicine, physics, and music. Understanding and managing the oscillatory motion for mechanical components, which has several practical applications, is made possible by vibration research. We shall examine a few of the typical uses of vibrations in this post (Vakakis, 1997).

6.14.1. Structural Health Monitoring

Monitoring the health of structures, including buildings, airplanes, and bridges, requires using vibration signals. A structure would vibrate at its native frequencies while exposed to outside pressures like wind and seismic activity. Researchers may identify abnormalities or faults, such as fractures or deformation, by examining the frequency and vibration patterns.

It enables engineers to locate possible weak places and act before a disastrous failure happens. Infrastructure security and dependability depend on vibration analysis-based structural health monitoring (Farrar & Worden, 2007).

6.14.2. Machine Condition Monitoring

Vibration analysis is frequently used to track equipment health, including motors, turbines, and pumps. Each machine generates a distinctive vibration pattern that could be examined to find any potential flaws or malfunctions. For instance, imbalance or misalignment might result in excessive vibration, which can hasten part failure and wear. Engineers can arrange repairs and repairs before such a machine break by monitoring the vibration patterns to find these flaws early (Peng & Chu, 2004).

6.14.3. Seismic Analysis

The study of how structures respond to seismic forces, including earthquakes, is known as the seismic response. Engineers utilize vibration analysis as a vital technique for seismic analysis because that enables them to predict how buildings will react to earthquake forces. Engineers could forecast how such a structure will react during an earthquake and build it to resist the effects of seismic activity by examining the structure's inherent frequencies and patterns of vibrations. In earthquake-prone areas, seismic analysis employing vibration analysis is essential for guaranteeing the security and dependability of buildings (Russell et al., 1997).

6.14.4. Ultrasonic Testing

A non-destructive technique called ultrasonic testing is utilized to find cracks and imperfections in materials like composites and metals. The material is probed using ultrasonic vibrations, and also the material's reaction is examined to look for any potential flaws or irregularities. Engineers may identify fractures, voids, or other flaws that can jeopardize the material's integrity by examining the vibration reaction of the material (Honarvar & Varvani, 2020).

6.14.5. Musical Instruments

Guitars, keyboards, and drums, among other musical instruments, depending on vibration to produce sound. A thread or membrane vibrates at its inherent frequency after being pulled or impacted, creating sound waves. Instrument designers may use the concepts of vibrations and resonance to create instruments that generate certain frequencies and timbres. For instance, a guitar string's thickness and tension affect its natural frequency, affecting the pitch of a note it produces (Millett & Randles, 1986).

6.14.6. Medical Imaging

Create pictures of the body's interior; vibrations are employed in medical imaging techniques like MRI and ultrasound. During ultrasonic imaging, a transducer vibrates at a certain frequency, and the tissue's reaction is examined to create pictures. In MRI, the protons within the body's tissues are vibrated by powerful magnetic fields, as well as the reaction of a proton is evaluated to create pictures. To detect and cure medical disorders, clinicians can create precise pictures of the inside body structures by vibrating its tissue and evaluating the reaction (Hounsfield, 1980).

6.14.7. Aerospace and Defense

Vibration analysis is widely employed in defense and aerospace applications, including missiles and spacecraft. Extreme pressures are applied to these systems and must operate dependably in challenging circumstances. Vibration analysis guarantees that the systems won't malfunction due to excessive vibration or wear and tear. Engineers may create these systems to survive the stresses of launching and spaceflight by studying the vibration of such systems, assuring their safe and dependable functioning (Srinivasan et al., 2021).

ACTIVITY 6.1.

What are the natural vibration frequencies and modes of a suspension bridge under wind loading, and how can this information be used to optimize the bridge's design and ensure its safety and durability over time?

6.15. SUMMARY

The chapter on Mechanical Vibrations covers the concept of mechanical vibrations in detail. It begins by explaining the mechanism of simple harmonic motion and how it is related to vibrations. It then introduces the concept of single degree of freedom systems (SDOF), which are the basic building blocks for understanding more complex vibrations. The chapter explains free vibration without damping and with viscous damping, which is an essential concept for understanding the behavior of vibrating systems. The concept of forced vibration of two degrees and multi-DOF systems is also discussed. The chapter highlights the importance of vibration control and suppression in various fields and explains how vibrations can be controlled and suppressed. Finally, the chapter explores the applications of vibrations in various fields such as seismic analysis of buildings, vibration testing of measuring instruments, and noise and vibration control in vehicles. It also emphasizes the significance of energy dissipation and energy methods in understanding the behavior of vibrating systems.

REVIEW QUESTIONS

1. What is the mechanism of simple harmonic motion?
2. What is a single degree of freedom system, and how does it vibrate?
3. What is free vibration, and how does it differ from forced vibration?
4. What is the significance of vibration control and suppression?
5. What are the applications of vibrations in various fields?
6. What is damping in vibrations, and how does it affect the vibration response?
7. What is a multi-degree of freedom system, and how does it vibrate?
8. How can the vibration response of a system be analyzed and controlled?

MULTIPLE CHOICE QUESTIONS

1. What is simple harmonic motion?

a. A motion that is non-periodic
b. A motion that is periodic and follows a sine or cosine function
c. A motion that is random
d. A motion that is straight-line

2. What is a single degree of freedom system?

a. A system that has only one type of motion
b. A system that has only one degree of freedom
c. A system that has only one component
d. A system that has only one type of force

3. **What is free vibration?**
   a. A vibration that is forced by an external force
   b. A vibration that is damped
   c. A vibration that occurs without any external force
   d. A vibration that is periodic
4. **What is forced vibration?**
   a. A vibration that is caused by an external force
   b. A vibration that is damped
   c. A vibration that occurs without any external force
   d. A vibration that is periodic
5. **What is the significance of vibration control?**
   a. To reduce noise levels
   b. To increase the life of mechanical components
   c. To improve the performance of mechanical systems
   d. All of the above
6. **What are the applications of vibrations?**
   a. Seismic analysis of buildings
   b. Measuring instruments for vibration testing
   c. Noise and vibration control in vehicles
   d. All of the above
7. What is damping in vibrations?
   a. The rate at which energy is lost in a vibrating system
   b. The frequency of vibration
   c. The amplitude of vibration
   d. The phase angle of vibration
8. **What is a multi-degree of freedom system?**
   a. A system that has more than one type of motion
   b. A system that has more than one degree of freedom
   c. A system that has more than one component
   d. A system that has more than one type of force

Answers to Multiple Choice Questions

1. (b); 2. (b); 3. (c); 4. (a); 5. (d); 6. (d); 7. (a); 8. (b)

REFERENCES

1. Adcroft, A., (2011). The mythology of feedback. Higher Education Research & Development, 30(4), 405–419.
2. Aguado, B. A., Grim, J. C., Rosales, A. M., Watson-Capps, J. J., & Anseth, K. S., (2018). Engineering precision biomaterials for personalized medicine. Science Translational Medicine, 10(424), eaam8645.
3. Akhras, G., (2000). Smart materials and smart systems for the future. Canadian Military Journal, 1(3), 25–31.
4. Alexiadou, A., & Doron, E., (2012). The syntactic construction of two non-active voices: Passive and middle. Journal of Linguistics, 48(1), 1–34.
5. Alkhatib, R., & Golnaraghi, M. F., (2003). Active structural vibration control: A review. Shock and Vibration Digest, 35(5), 367.
6. Balaji, P. S., & SelvaKumar, K. K., (2021). Applications of nonlinearity in passive vibration control: A review. Journal of Vibration Engineering & Technologies, 9, 183–213.
7. Baltzopoulos, G., Baraschino, R., Iervolino, I., & Vamvatsikos, D., (2018). Dynamic analysis of single-degree-of-freedom systems (DYANAS): A graphical user interface for OpenSees. Engineering Structures, 177, 395–408.
8. Bapat, C. N., & Sankar, S., (1985). Single unit impact damper in free and forced vibration. Journal of Sound and Vibration, 99(1), 85–94.
9. Basili, M., De Angelis, M., & Pietrosanti, D., (2019). Defective two adjacent single degree of freedom systems linked by spring-dashpot-inerter for vibration control. Engineering Structures, 188, 480–492.
10. Benware, C. A., & Deci, E. L., (1984). Quality of learning with an active versus passive motivational set. American Educational Research Journal, 21(4), 755–765.
11. Blazejczyk-Okolewska, B., Czolczynski, K., & Kapitaniak, T., (2009). Dynamics of a two-degree-of-freedom cantilever beam with impacts. Chaos, Solitons & Fractals, 40(4), 1991–2006.
12. Blum, E. K., & Li, L. K., (1991). Approximation theory and feedforward networks. Neural Networks, 4(4), 511–515.
13. Brennan, M. J., (1997). Vibration control using a tunable vibration neutralizer. Proceedings of the Institution of Mechanical Engineers, Part C: Journal of Mechanical Engineering Science, 211(2), 91–108.
14. Brody, T. A., Flores, J., French, J. B., Mello, P. A., Pandey, A., & Wong, S. S., (1981). Random-matrix physics: Spectrum and strength fluctuations. Reviews of Modern Physics, 53(3), 385.
15. Cao, B., & Sarkar, P. P., (2012). Identification of rational functions using two-degree-of-freedom model by forced vibration method. Engineering Structures, 43, 21–30.
16. Chaudet, C., Fleury, E., Lassous, I. G., Rivano, H., & Voge, M. E., (2005). Optimal

positioning of active and passive monitoring devices. In: Proceedings of the 2005 ACM Conference on Emerging Network Experiment and Technology (pp. 71–82).

17. Chen, J. S., Chen, J. Y., & Chou, Y. F., (2008). On the natural frequencies and mode shapes of dragonfly wings. Journal of Sound and Vibration, 313(3–5), 643–654.
18. Cheng, C. C., & Wang, J. Y., (2003). Free vibration analysis of a resilient impact damper. International Journal of Mechanical Sciences, 45(4), 589–604.
19. Choi, K. M., Jung, H. J., Cho, S. W., & Lee, I. W., (2007). Application of smart passive damping system using MR damper to highway bridge structure. Journal of Mechanical Science and Technology, 21, 870–874.
20. Chrastil, E. R., & Warren, W. H., (2012). Active and passive contributions to spatial learning. Psychonomic Bulletin & Review, 19, 1–23.
21. Christopoulos, C., Pampanin, S., & Nigel, P. M. J., (2003). Performance-based seismic response of frame structures including residual deformations part I: Single-degree of freedom systems. Journal of Earthquake Engineering, 7(01), 97–118.
22. Coron, J. M., & D'Andrea-Novel, B., (1998). Stabilization of a rotating body beam without damping. IEEE Transactions on Automatic Control, 43(5), 608–618.
23. Cosenza, E., Manfredi, G., & Polese, M., (2009). Simplified method to include cumulative damage in the seismic response of single-degree-of-freedom systems. Journal of Engineering Mechanics, 135(10), 1081–1088.
24. Crago, P. E., Lan, N., Veltink, P. H., Abbas, J. J., & Kantor, C., (1996). New control strategies for neuroprosthetic systems. Journal of Rehabilitation Research and Development, 33, 158–172.
25. Dahl, J. M., Hover, F. S., & Triantafyllou, M. S., (2006). Two-degree-of-freedom vortex-induced vibrations using a force assisted apparatus. Journal of Fluids and Structures, 22(6, 7), 807–818.
26. Daley, S., Johnson, F. A., Pearson, J. B., & Dixon, R., (2004). Active vibration control for marine applications. Control Engineering Practice, 12(4), 465–474.
27. Deak, G., Curran, K., & Condell, J., (2012). A survey of active and passive indoor localization systems. Computer Communications, 35(16), 1939–1954.
28. Desmond, P. A., & Hancock, P. A., (2000). Active and passive fatigue states. In: Stress, Workload, and Fatigue (pp. 455–465). CRC Press.
29. Eldan, R., & Shamir, O., (2016). The power of depth for feedforward neural networks. In: Conference on Learning Theory (pp. 907–940). PMLR.
30. Farrar, C. R., & Worden, K., (2007). An introduction to structural health monitoring. Philosophical Transactions of the Royal Society A: Mathematical, Physical and Engineering Sciences, 365(1851), 303–315.
31. Feldman, M., (1994). Non-linear system vibration analysis using Hilbert transform--II. Forced vibration analysis method 'Forcevib.' Mechanical Systems and Signal

Processing, 8(3), 309–318.
32. Feldman, M., (1997). Non-linear free vibration identification via the Hilbert transform. Journal of Sound and Vibration, 208(3), 475–489.
33. Findeisen, D., & Popp, K., (2002). Systems dynamics and mechanical vibrations: An introduction. Appl. Mech. Rev., 55(3), B49, B50.
34. Finez, A., Pereira, A., & Leclère, Q., (2015). Broadband mode decomposition of ducted fan noise using cross-spectral matrix denoising. In: FAN2015 Conference (pp. 15–17).
35. Forsythe, A., & Johnson, S., (2017). Thanks, but no-thanks for the feedback. Assessment & Eva.
36. Franchek, M. A., Ryan, M. W., & Bernhard, R. J., (1996). Adaptive passive vibration control. Journal of Sound and Vibration, 189(5), 565–585.
37. Gabbert, U., (2002). Research activities in smart materials and structures and expectations to future developments. Journal of Theoretical and Applied Mechanics, 40(3), 549–574.
38. Galeriu, C., Edwards, S., & Esper, G., (2014). An Arduino investigation of simple harmonic motion. The Physics Teacher, 52(3), 157–159.
39. Gao, X., Tracy, L. A., Anderson, E. M., Campbell, D. M., Ivie, J. A., Lu, T. M., & Misra, S., (2020). Modeling assisted room temperature operation of atomic precision advanced manufacturing devices. In: 2020 International Conference on Simulation of Semiconductor Processes and Devices (SISPAD) (pp. 277–280). IEEE.
40. Gilar, M., Olivova, P., Daly, A. E., & Gebler, J. C., (2005). Orthogonality of separation in two-dimensional liquid chromatography. Analytical Chemistry, 77(19), 6426–6434.
41. Glorot, X., & Bengio, Y., (2010). Understanding the difficulty of training deep feedforward neural networks. In: Proceedings of the Thirteenth International Conference on Artificial Intelligence and Statistics (pp. 249–256). JMLR Workshop and Conference Proceedings.
42. Greller, M. M., & Herold, D. M., (1975). Sources of feedback: A preliminary investigation. Organizational Behavior and Human Performance, 13(2), 244–256.
43. Gsell, S., Bourguet, R., & Braza, M., (2016). Two-degree-of-freedom vortex-induced vibrations of a circular cylinder at Re= 3900. Journal of Fluids and Structures, 67, 156–172.
44. Habib, G., Cirillo, G. I., & Kerschen, G., (2017). Uncovering detached resonance curves in single-degree-of-freedom systems. Procedia Engineering, 199, 649–656.
45. Haddow, A. G., Barr, A. D. S., & Mook, D. T., (1984). Theoretical and experimental study of modal interaction in a two-degree-of-freedom structure. Journal of Sound and Vibration, 97(3), 451–473.
46. Harvey, L., (2003). Student feedback [1]. Quality in Higher Education, 9(1), 3–20.
47. Hattie, J., & Timperley, H., (2007). The power of feedback. Review of Educational

Research, 77(1), 81–112.

48. He, W., & Ge, S. S., (2015). Vibration control of a flexible beam with output constraint. IEEE Transactions on Industrial Electronics, 62(8), 5023–5030.
49. Held, I. M., (1985). Pseudomomentum and the orthogonality of modes in shear flows. Journal of Atmospheric Sciences, 42(21), 2280–2288.
50. Honarvar, F., & Varvani-Farahani, A., (2020). A review of ultrasonic testing applications in additive manufacturing: Defect evaluation, material characterization, and process control. Ultrasonics, 108, 106227.
51. Hornik, K., (1991). Approximation capabilities of multilayer feedforward networks. Neural Networks, 4(2), 251–257.
52. Hornik, K., Stinchcombe, M., & White, H., (1989). Multilayer feedforward networks are universal approximators. Neural Networks, 2(5), 359–366.
53. Hounsfield, G. N., (1980). Computed medical imaging. Science, 210(4465), 22–28.
54. Hung, J. M., Huang, Y. H., Huang, S. P., Chang, F. C., Wen, T. H., Su, C. I., & Chang, M. F., (2022). An 8-Mb DC-current-free binary-to-8b precision ReRAM nonvolatile computing-in-memory macro using time-space-readout with 1286.4-21.6 TOPS/W for edge-AI devices. In: 2022 IEEE International Solid-State Circuits Conference (ISSCC) (Vol. 65, pp. 1–3). IEEE.
55. Hung, J. M., Xue, C. X., Kao, H. Y., Huang, Y. H., Chang, F. C., Huang, S. P., & Chang, M. F., (2021). A four-megabit compute-in-memory macro with eight-bit precision based on CMOS and resistive random-access memory for AI edge devices. Nature Electronics, 4(12), 921–930.
56. Ilgen, D. R., Fisher, C. D., & Taylor, M. S., (1979). Consequences of individual feedback on behavior in organizations. Journal of Applied Psychology, 64(4), 349.
57. Jafari, N., & Azhari, M., (2021). Free vibration analysis of viscoelastic plates with simultaneous calculation of natural frequency and viscous damping. Mathematics and Computers in Simulation, 185, 646–659.
58. Jalili, N., (2002). A comparative study and analysis of semi-active vibration-control systems. J. Vib. Acoust., 124(4), 593–605.
59. Jauvtis, N., & Williamson, C. H. K., (2003). Vortex-induced vibration of a cylinder with two degrees of freedom. Journal of Fluids and Structures, 17(7), 1035–1042.
60. Jeon, D., & Gharib, M., (2001). On circular cylinders undergoing two-degree-of-freedom forced motions. Journal of Fluids and Structures, 15(3, 4), 533–541.
61. Kazdaridis, G., Zografopoulos, I., Sidiropoulos, N., Symeonidis, P., & Korakis, T., (2020). eProfiler: High-precision power monitoring system for IoT devices featuring extreme dynamic range of operation. In: Proceedings of the 8th International Workshop on Energy Harvesting and Energy-Neutral Sensing Systems (pp. 50–56).
62. Kinchin, J., (2016). Using tracker to prove the simple harmonic motion equation. Physics Education, 51(5), 053003.

63. Kluger, A. N., & Nir, D., (2010). The feedforward interview. Human Resource Management Review, 20(3), 235–246.
64. Kolcaba, K. Y., & Kolcaba, R. J., (1991). An analysis of the concept of comfort. Journal of Advanced Nursing, 16(11), 1301–1310.
65. Krishna, B. T., & Reddy, K. V. V. S., (2008). Active and passive realization of fractance device of order 1/2. Active and Passive Electronic Components, 2008.
66. Lamme, V. A., & Roelfsema, P. R., (2000). The distinct modes of vision offered by feedforward and recurrent processing. Trends in Neurosciences, 23(11), 571–579.
67. Lan, K. C., & Shih, W. Y., (2012). Using simple harmonic motion to estimate walking distance for waist-mounted PDR. In: 2012 IEEE Wireless Communications and Networking Conference (WCNC) (pp. 2445–2450). IEEE.
68. Lancaster, P., (1960). Free vibration and hysteretic damping. The Aeronautical Journal, 64(592), 229–229.
69. Leardini, A., O'connor, J. J., Catani, F., & Giannini, S., (1999). Kinematics of the human ankle complex in passive flexion; a single degree of freedom system. Journal of Biomechanics, 32(2), 111–118.
70. Liang, J. W., & Feeny, B. F., (1998). Identifying Coulomb and viscous friction from free-vibration decrements. Non-linear Dynamics, 16, 337–347.
71. Liang, J. W., (2005). Identifying Coulomb and viscous damping from free-vibration acceleration decrements. Journal of Sound and Vibration, 282(3–5), 1208–1220.
72. Liu, K., Tang, W., Zhou, F., & Qiu, G., (2019). Spectral regularization for combating mode collapse in Gans. In: Proceedings of the IEEE/CVF International Conference on Computer Vision (pp. 6382–6390).
73. Mackerle, J., (1998). Smart materials and structures-a finite-element approach: A bibliography (1986–1997). Modelling and Simulation in Materials Science and Engineering, 6(3), 293.
74. Mackerle, J., (2003). Smart materials and structures—A finite element approach—An addendum: A bibliography (1997–2002). Modelling and Simulation in Materials Science and Engineering, 11(5), 707.
75. Marino, L., & Cicirello, A., (2020). Experimental investigation of a single-degree-of-freedom system with Coulomb friction. Non-linear Dynamics, 99(3), 1781–1799.
76. Mead, D., (1975). Wave propagation and natural modes in periodic systems: II. Multi-coupled systems, with and without damping. Journal of Sound and Vibration, 40(1), 19–39.
77. Mehrabi, A. B., & Tabatabai, H., (1998). Unified finite difference formulation for free vibration of cables. Journal of Structural Engineering, 124(11), 1313–1322.
78. Mestas Nuñez, A. M., (2000). Orthogonality properties of rotated empirical modes. International Journal of Climatology: A Journal of the Royal Meteorological Society, 20(12), 1509–1516.
79. Millett, S. M., & Randles, F., (1986). Scenarios for strategic business planning: A

case history for aerospace and defense companies. Interfaces, 16(6), 64–72.

80. Mohan, V., Reddy, Y. J., & Kalpana, K., (2011). Active and passive network measurements: A survey. International Journal of Computer Science and Information Technologies, 2(4), 1372–1385.
81. Nayfeh, (1985). The response of non-linear single-degree-of-freedom systems to multifrequency excitations. Journal of Sound and Vibration, 102(3), 403–414.
82. Nayfeh, A. H., (1983). The response of single degree of freedom systems with quadratic and cubic non-linearities to a subharmonic excitation. Journal of Sound and Vibration, 89(4), 457–470.
83. Nayfeh, A. H., Kreider, W., & Anderson, T. J., (1995). Investigation of natural frequencies and mode shapes of buckled beams. AIAA Journal, 33(6), 1121–1126.
84. Olmos, B. A., & Roesset, J. M., (2010). Evaluation of the half power bandwidth method to estimate damping in systems without real modes. Earthquake Engineering & Structural Dynamics, 39(14), 1671–1686.
85. Onoda, J., Endo, T., Tamaoki, H., & Watanabe, N., (1991). Vibration suppression by variable-stiffness members. AIAA Journal, 29(6), 977–983.
86. Fraser, W. B. (1976). Orthogonality relation for the Rayleigh–Lamb modes of vibration of a plate. The Journal of the Acoustical Society of America, 59(1), 215-216.
87. Parker, K. J., Huang, S. R., Musulin, R. A., & Lerner, R. M., (1990). Tissue response to mechanical vibrations for "sonoelasticity imaging." Ultrasound in Medicine & Biology, 16(3), 241–246.
88. Peng, Z. K., & Chu, F. L., (2004). Application of the wavelet transform in machine condition monitoring and fault diagnostics: A review with bibliography. Mechanical Systems and Signal Processing, 18(2), 199–221.
89. Pimental, G. C., & Charles, S. W., (1963). Infrared spectral perturbations in matrix experiments. Pure and Applied Chemistry, 7(1), 111–124.
90. Pinto, J. K., & Mantel, S. J., (1990). The causes of project failure. IEEE Transactions on Engineering Management, 37(4), 269–276.
91. Rachels, J., (1975). Active and passive euthanasia. Bioethics: An Introduction to the History, Methods, and Practice, 77–82.
92. Rao, S. S., & Yap, F. F., (1995). Mechanical Vibrations (Vol. 4, pp. 75–848). New York: Addison-Wesley.
93. Rao, S. S., (1974). Natural vibrations of systems of elastically connected Timoshenko beams. The Journal of the Acoustical Society of America, 55(6), 1232–1237.
94. Reid, T. J., (1956). Free vibration and hysteretic damping. The Aeronautical Journal, 60(544), 283–285.
95. Rogelberg, S. G., Conway, J. M., Sederburg, M. E., Spitzmüller, C., Aziz, S., & Knight, W. E., (2003). Profiling active and passive nonrespondents to an organizational survey. Journal of Applied Psychology, 88(6), 1104.

96. Russell, B., Hampson, D., Schuelke, J., & Quirein, J., (1997). Multiattribute seismic analysis. The Leading Edge, 16(10), 1439–1444.
97. Sadeghipour, K., Salomon, R., & Neogi, S., (1992). Development of a novel electrochemically active membrane and 'smart' material based vibration sensor/damper. Smart Materials and Structures, 1(2), 172.
98. Salawu, O. S., & Williams, C., (1995). Bridge assessment using forced-vibration testing. Journal of Structural Engineering, 121(2), 161–173.
99. Sanders, B., Crowe, R., & Garcia, E., (2004). Defense advanced research projects agency–smart materials and structures demonstration program overview. Journal of Intelligent Material Systems and Structures, 15(4), 227–233.
100. Seidler, R. D., Noll, D. C., & Thiers, G., (2004). Feedforward and feedback processes in motor control. Neuroimage, 22(4), 1775–1783.
101. Somroob, S., & Wattanakasiwich, P., (2017). Investigating student understanding of simple harmonic motion. In: Journal of Physics: Conference Series (Vol. 901, No. 1, p. 012123). IOP Publishing.
102. Spillman, Jr. W. B., Sirkis, J. S., & Gardiner, P. T., (1996). Smart materials and structures: What are they? Smart Materials and Structures, 5(3), 247.
103. Srinivasan, V., Kunjiappan, S., & Palanisamy, P., (2021). A brief review of carbon nanotube reinforced metal matrix composites for aerospace and defense applications. International Nano Letters, 1–25.
104. Stanway, R., Sproston, J. L., & El-Wahed, A. K., (1996). Applications of electro-rheological fluids in vibration control: A survey. Smart Materials and Structures, 5(4), 464.
105. Sun, J., Guan, Q., Liu, Y., & Leng, J., (2016). Morphing aircraft based on smart materials and structures: A state-of-the-art review. Journal of Intelligent Material Systems and Structures, 27(17), 2289–2312.
106. Thenozhi, S., & Yu, W., (2013). Advances in modeling and vibration control of building structures. Annual Reviews in Control, 37(2), 346–364.
107. Triana, C. A., & Fajardo, F., (2011). The influence of spring length on the physical parameters of simple harmonic motion. European Journal of Physics, 33(1), 219.
108. Trifunac, M. D., (1972). Comparisons between ambient and forced vibration experiments. Earthquake Engineering & Structural Dynamics, 1(2), 133–150.
109. Udwadia, F. E., & Tabaie, S., (1981). Pulse control of single-degree-of-freedom system. Journal of the Engineering Mechanics Division, 107(6), 997–1009.
110. Vakakis, A., (1997). Non-linear normal modes (NNMs) and their applications in vibration theory: An overview. Mechanical Systems and Signal Processing, 11(1), 3–22.
111. Van, R. N. F. J., & Van, D. M. A., (2006). Natural frequencies and modes of a Timoshenko beam. Wave Motion, 44(1), 58–69.
112. Varanis, M., Silva, A., Mereles, A., & Pederiva, R., (2018). MEMS accelerometers for mechanical vibrations analysis: A comprehensive review with applications. Journal of

the Brazilian Society of Mechanical Sciences and Engineering, 40, 1–18.

113. Villaverde, R., & Koyama, L. A., (1993). Damped resonant appendages to increase inherent damping in buildings. Earthquake Engineering & Structural Dynamics, 22(6), 491–507.

114. Vinyas, M., (2021). Computational analysis of smart magneto-electro-elastic materials and structures: Review and classification. Archives of Computational Methods in Engineering, 28(3), 1205–1248.

115. Xue, C. X., Hung, J. M., Kao, H. Y., Huang, Y. H., Huang, S. P., Chang, F. C., & Chang, M. F., (2021). 16.1 A 22nm 4Mb 8b-precision ReRAM computing-in-memory macro with 11.91 to 195.7 TOPS/W for tiny AI edge devices. In: 2021 IEEE International Solid-State Circuits Conference (ISSCC) (Vol. 64, pp. 245–247). IEEE.

116. Yamakoshi, Y., Sato, J., & Sato, T., (1990). Ultrasonic imaging of internal vibration of soft tissue under forced vibration. IEEE Transactions on Ultrasonics, Ferroelectrics, and Frequency Control, 37(2), 45–53.

117. Yeh, G. C., (1966). Forced vibrations of a two degree of freedom system with combined coulomb and viscous damping. The Journal of the Acoustical Society of America, 39(1), 14–24.

118. Zhang, B. L., Han, Q. L., & Zhang, X. M., (2017). Recent advances in vibration control of offshore platforms. Non-linear Dynamics, 89, 755–771.

119. Zhao, Z., Niu, J., Shen, Y., & Yang, S., (2021). Forced vibration of two-degrees-of-freedom machine tool feed system with clearance and friction. Applied Mathematical Modeling, 92, 281–296.

CHAPTER 7

SIMPLE MACHINES

UNIT INTRODUCTION

The majority of basic machines consist of a single moving element. Combining simple machines yields sophisticated machines, such as a lawnmower, a car, or even an electronic nose hair trimmer! A machine is described as every tool that facilitates work. In science, "labor" means to cause motion. It is essential to understand that when you operate a basic machine, you are actually performing the very same amount of work; it only appears easier.

So, there is an energy trade-off when using small machines in science. "Work" is only performed when an object is moved. For instance, pushing on a wall does not constitute work because it cannot be moved. The work consists of two components. The first is the quantity of force (push or pull) required to complete the task. The distance across whereby the force is exerted is the second factor. Work is defined by the following formula (Insall & Machesky, 2009):

Work = Force × Distance

Force is the pull or push on an object that causes it to move. Distance is the distance traveled by an object. Hence, the work accomplished is equal to the force exerted times the distance moved.

When we say that a machine makes work easier, we imply that it requires less effort to perform the same amount of labor. In addition to allowing us to increase the distance over which a smaller force is delivered, machines may also allow us to change the direction of a force that is applied. Although machines do not lower our workload, they can make things easier (Marulcu & Barnett, 2016) (Figure 7.1).

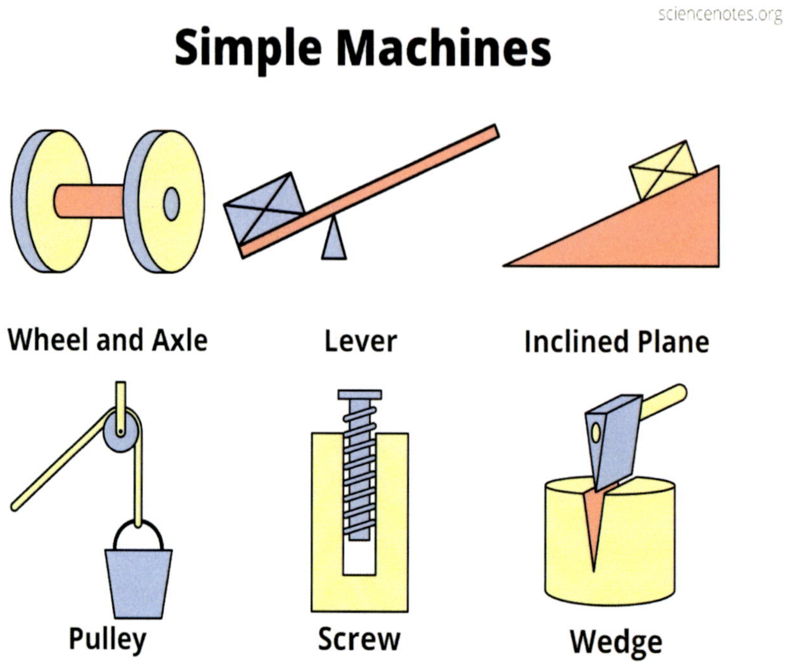

Figure 7.1. Illustration of simple machines.

Source: Direct Science, Creative Commons License.

Learning Objectives

At the end of this chapter, readers will be able to:

- Describe the different types of simple machines (e.g., pulleys, inclined planes, levers, wheels and axles, wedges, and screws).
- Explain the principles behind how each type of simple machine works.
- Identify the advantages and disadvantages of using simple machines in various applications.
- Analyze the mechanical advantage of a simple machine.
- Understand the concept of work and power in relation to simple machines.
- Evaluate the efficiency of simple machines in practical situations.
- Recognize the role of simple machines in modern technology.
- Apply knowledge of simple machines to solve practical problems.

Key Terms

- Applications
- Inclined plane
- Lever
- Machines
- Pulley
- Screw
- Transportation
- Wedge
- Wheel and axle

7.1. TYPES OF SIMPLE MACHINES

Understanding the six types of simple machines is important because they are the building blocks of more complex machinery. By understanding how these simple machines work, we can gain a better understanding of the world around us and the technology that we use every day (Norbury, 2006).

1. **The Lever:** A lever is a bar that pivots on a fulcrum to lift or move objects. Levers are used in many different applications, from construction equipment to scissors and seesaws (Chickering & Ehrmann, 1996).
2. **The Pulley:** A pulley is a wheel with a grooved rim around its edge, which is used to change the direction or magnitude of a force. Pulleys are commonly used in cranes and other lifting equipment (Zhu et al., 2010).
3. **The Wheel and Axle:** A wheel and axle is a simple machine consisting of a wheel attached to a shaft or axle. This simple machine is used to reduce friction and increase speed in many different applications, from transportation to manufacturing (Arduini et al., 2008).
4. **The Inclined Plane:** An inclined plane is a sloping surface that allows objects to be raised or lowered with less force than would be required to lift them directly. Examples of inclined planes include ramps and staircases (Wartman et al., 2003).
5. **The Wedge:** A wedge is a simple machine that is used to separate two objects or to hold an object in place. Examples of wedges include knives, axes, and doorstops (Cognard, 1986).
6. **The Screw:** A screw is a simple machine that consists of a shaft with a spiral groove around it. Screws are commonly used to hold objects together or to lift objects vertically (Rorres, 2000).

Remember

A simple machine reduces the amount of effort needed to move something, but you wind up moving it a greater distance to accomplish the same amount of work.

7.2. PULLEY

The primary function of a pulley is to alter the orientation of an applied force, hence decreasing the quantity of effort and force needed to shift an object. By exerting a downward force to a pulley, an object will be raised. Loop the thread and hook of the 10 g block above one pulley wheel and pull down on the hook to illustrate this principle. See how the block rises upwards when the hook is dragged downwards. The imparted force alters the block's movement direction, making it easier to ascend. Picture a construction worker attempting to lift a huge beam to the building's peak. A machine with a pulley system would make it simpler to raise the beam upward (Demer, 2002).

A pulley is a chain or wire that passes over at least one wheel or a series of wheels. Pulleys can be seen on flagpoles, construction cranes, window blinds, and vintage elevators. Play with the pulley model by altering the placement, number, and size of the wheels. Put washers to the hook's end. How many washers are required to move the 5 g and 10 g blocks? Does effort vary when the string passes through more or fewer wheels? How does the course alter? Increase or reduction in effort while utilizing small or large wheels? Does effort vary if the position of the wheels is altered? How does the course change? (Kong & Parker, 2005) (Figure 7.2).

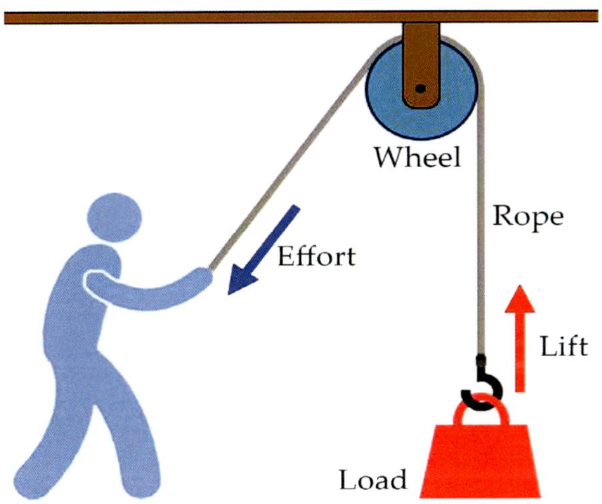

Figure 7.2. Diagram of working of pulley.

Source: Science Facts, Creative Commons License.

TIPS

Pulleys were used in ancient times to lift heavy objects like water buckets. So, the next time you need to lift something heavy, try using a pulley system and save yourself some muscle power.

7.3. INCLINED PLANE

An inclined plane is a simple machine that consists of a flat surface that is sloped at an angle. It is used to move objects from one level to another by reducing the amount of force needed to lift them. When an object is moved up an inclined plane, it travels a longer distance than if it were lifted straight up, but it requires less force to do so. This is because the force required to move an object up an inclined plane is spread out over a longer distance, making it easier to push or pull the object up the slope (Yih, 1963).

There are several types of inclined planes, including ramps, chutes, and screws. Ramps are the most common type of inclined plane and are used to move heavy objects up and down stairs, loading docks, and other elevated surfaces. Chutes are inclined planes that are used to guide objects or materials from a higher level to a lower level, such as a grain chute or a waste disposal chute. Screws are inclined planes that are wrapped around a cylinder and are used to lift or move objects, such as the screw on a vise or a bottle cap (Benjamin, 1957).

The mechanical advantage of an inclined plane is determined by the angle of the slope. The steeper the slope, the greater the mechanical advantage, but the more force is required to push or pull the object up the slope. The shallower the slope, the smaller the mechanical advantage, but the less force is required to move the object up the slope. The mechanical advantage of an inclined plane is calculated by dividing the length of the slope by the height of the slope. This is known as the slope angle (Lister, 1992). For example, if an inclined plane has a slope length of 4 meters and a slope height of 1 meter, the slope angle would be 4:1. This means that the mechanical advantage of the inclined plane is 4, which means that it reduces the amount of force required to move an object by a factor of 4. Inclined planes are commonly used in everyday life, such as in the form of ramps for wheelchairs and strollers. They are also used in industrial applications, such as conveyor belts and loading ramps for trucks. The inclined plane is a fundamental tool that has been used by humans for thousands of years, and its simple yet effective design makes it a valuable tool in many different settings (Pumir et al., 1983) (Figure 7.3).

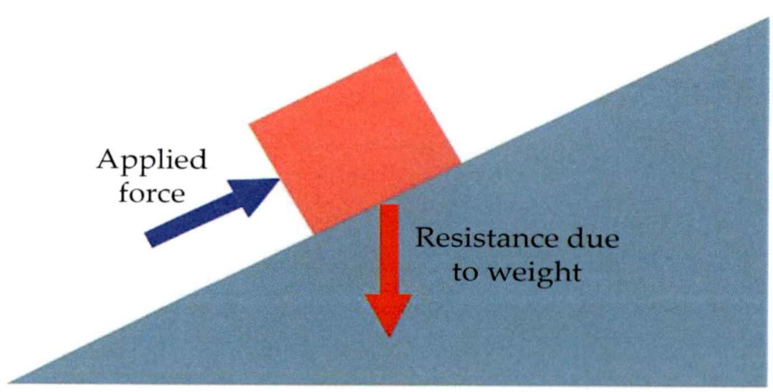

Figure 7.3. Schematic of inclined plane.

Source: Energy Education, Creative Commons License.

7.4. LEVER

A lever is a simple machine consisting of a rigid bar or plank that rotates around a fixed point, known as the fulcrum. Levers are used to lift or move heavy objects with less force than would be required without the machine. The lever works on the principle of force multiplication, which means that a small force can be used to lift a larger weight. The amount of force required to lift an object depends on the distance from the fulcrum to the object and the distance from the fulcrum to the point where the force is applied (Geeves, 2002).

There are three types of levers, depending on the position of the fulcrum, load, and effort. In a first-class lever, the fulcrum is located between the load and effort, like a seesaw. In a second-class lever, the load is located between the fulcrum and effort, like a wheelbarrow. In a third-class lever, the effort is located between the fulcrum and load, like a pair of tweezers. Each type of lever has unique advantages and disadvantages, depending on the task at hand (Sellars, 1981). Levers have been used by humans for thousands of years, from simple tools like pry bars to complex machines like cranes and catapults. They have played a vital role in shaping our civilization, enabling us to move heavy objects and perform tasks that would otherwise be impossible. Today, levers are used in a wide range of applications, from construction and manufacturing to sports and entertainment. For example, the pole vault in track and field is a classic example of a third-class lever, where the athlete uses a long pole to lift themselves over a high bar (Purcell et al., 2002) (Figure 7.4).

Figure 7.4. Images of examples of lever.

Source: Edinformatics, Creative Commons License.

7.5. WHEEL AND AXLE

A wheel and axle are a simple machine that consists of a circular wheel attached to a smaller cylindrical axle. The wheel rotates around the axle, which is fixed in place, and both parts move together. The wheel and axle are designed to reduce the amount of force needed to move an object by transferring the force over a greater distance. The larger the wheel, the greater the distance the force is transferred, and the easier it is to move the object (Soldatov, 2006).

7.5.1. Types of Wheel and Axle Systems

There are several types of wheel and axle systems, each with its unique features and applications. Here are some of the most common types of wheel and axle systems and examples of their use:

7.5.1.1. Windlass

A windlass is a type of wheel and axle system that is used to lift heavy objects vertically. It consists of a cylindrical axle with a crank handle attached to it and a circular wheel that is attached to a rope or cable. By turning the crank handle, the wheel rotates, and the rope or cable is wound around the axle, lifting the heavy object. Windlasses are commonly used for lifting buckets from wells, hoisting sails on ships, and in construction to lift heavy loads (Thordarson et al., 1997).

7.5.1.2. Gears

Gears are a type of wheel and axle system that is used to transfer motion and power between two or more wheels.

A gear consists of two or more wheels that are connected by teeth that mesh together. When one gear turns, it causes the other gears to turn as well, allowing for the transfer of motion and power. Gears are commonly used in bicycles, automobiles, clocks, and other machinery (Petrescu et al., 2017).

7.5.1.3. Cams

A cam is a type of wheel and axle system that is used to convert rotary motion into linear motion. It consists of a circular wheel with an irregular shape that is used to push or pull a lever or rod.

As the wheel rotates, the shape of the cam causes the lever or rod to move up and down or back and forth. Cams are commonly used in engines, pumps, and other machinery (Sadek & Daadbin, 1990).

7.5.1.4. Flywheel

A flywheel is a type of wheel and axle system that is used to store energy and smooth out the motion of a machine. It consists of a large circular wheel that is attached to a rotating shaft. As the machine operates, the flywheel stores energy in its rotational motion, which can be used to smooth out any variations in the machine's operation. Flywheels are commonly used in engines, generators, and other machinery (Bitterly, 1998) (Figure 7.5).

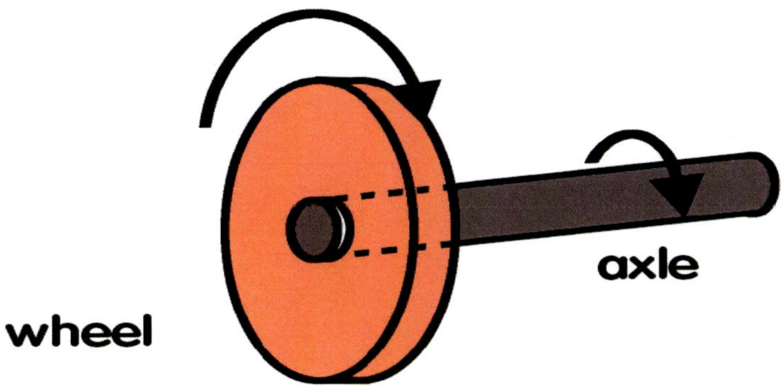

Figure 7.5. Picture of wheel and axle.

Source: Chalkboard, Creative Commons License.

7.5.2. Mechanical Advantage of a Wheel and Axle

The mechanical advantage of a wheel and axle is the ratio of the diameter of the wheel to the diameter of the axle. The larger the diameter of the wheel compared to the axle, the greater the mechanical advantage. The mechanical advantage of a wheel and axle is calculated by dividing the diameter of the wheel by the diameter of the axle (Singh et al., 2014).

The mechanical advantage of a wheel and axle can be explained by the principle of torque. Torque is the force that causes an object to rotate around an axis or pivot point. In the case of a wheel and axle, torque is created by the force applied to the rim of the wheel, which is then transferred to the axle through the spokes or shaft.

The larger the diameter of the wheel, the greater the torque that can be generated by a given force. This is because the force applied to the rim of the wheel is applied over a larger distance, resulting in a greater torque. The axle, on the other hand, is designed to transfer this torque to the load being moved (Veeger et al., 1991). One example of the mechanical advantage of a wheel and axle is a bicycle. The wheel of a bicycle is much larger than the axle, giving it a mechanical advantage that allows the rider to cover long distances with relatively little effort. The mechanical advantage of a bicycle wheel and axle system can be further increased by using gears, which allow the rider to adjust the ratio of the wheel to the axle depending on the terrain and desired speed (Vowles, 1932).

Another example of the mechanical advantage of a wheel and axle is a winch used to lift heavy loads. The diameter of the winch drum is much larger than the diameter of the winch shaft, giving it a mechanical advantage that allows it to lift heavy loads with relatively little effort. By using a system of pulleys, the mechanical advantage of the winch can be further increased, allowing it to lift even heavier loads (Nelson, 1980).

7.6. WEDGE

The wedge is a simple machine that is used to separate or lift objects apart. It is made up of two inclined planes joined together at the base. The pointed end of the wedge is pushed or driven into the material, causing it to split or lift apart. Wedges are commonly used in construction, woodworking, and agriculture. There are several types of wedges, including the simple wedge, the double wedge, and the blade wedge. The simple wedge is a straight, V-shaped piece of material that is used to split objects apart. The double wedge, also known as a chisel, has two sloping sides and a sharp edge that is used to cut materials. The blade wedge is a curved wedge that is used to cut through hard or tough materials (Brühwiler & Wittmann, 1990).

The mechanical advantage of a wedge is determined by the angle of the slope and the thickness of the wedge. The thinner the wedge and the steeper the angle of the slope, the greater the mechanical advantage. This is because a thin, steep wedge requires less force to be applied to it in order to split or lift an object. Wedges have been used since ancient times to perform a wide range of tasks. For example, stone wedges were used by the ancient Egyptians to split rocks for construction. The Romans used wedges to lift heavy objects such as statues and columns. In modern times, wedges are used in construction to split and shape stone, concrete, and other materials. Wedges are also used in agriculture. For example, plows use wedge-shaped blades to break up and turn over soil. Axes and hatchets use wedge-shaped heads to split wood for fuel and construction. In the food industry, knives, and cleavers use wedge-shaped blades to cut and separate meat, vegetables, and other food items.

In conclusion, the wedge is an important simple machine that is used to split and lift objects apart. Its mechanical advantage is determined by the angle of the slope and the thickness of the wedge. Wedges have been used since ancient times and are still used in a wide range of industries, including construction, agriculture, and the food industry (Cumberbatch, 1960) (Figure 7.6).

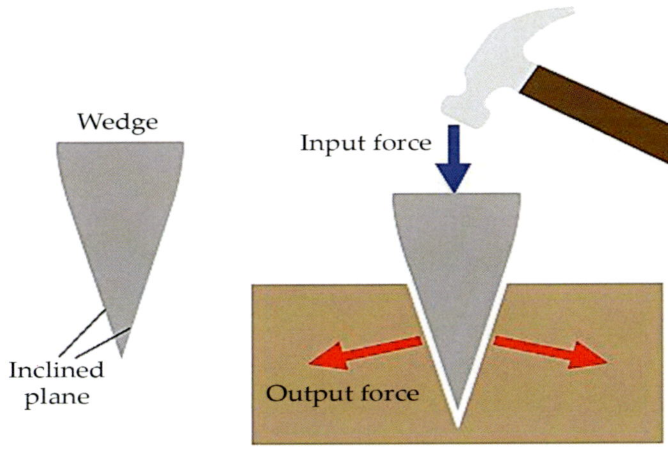

Figure 7.6. Illustration of wedge.

Source: Eschool Today, Creative Commons License.

7.7. SCREW

The screw is a simple machine that consists of a threaded shaft (or "screw") and a head that can be turned to drive the shaft into or out of a material. The shaft is typically pointed and has a spiral groove (known as the "thread") that runs along its length. Screws are commonly used to hold objects together, lift objects vertically, or to apply pressure in order to move objects.

One of the key benefits of screws is that they allow for a lot of force to be applied to a small area. This is because the force applied to the screw head is distributed along the length of the thread, which increases the amount of friction between the screw and the material it is being driven into. This friction provides a strong hold that can be used to secure objects together or to lift objects vertically (DeRosier, 1998).

There are many different types of screws, each with their own specific design and use. For example, wood screws are designed with a sharp point that makes it easy to drive them into wood, while machine screws are designed to be used with nuts and threaded holes. Screws can also be designed with different thread patterns and sizes, which can affect their holding strength and the amount of torque required to turn them.

One of the most common uses for screws is in woodworking. Wood screws are used to hold boards together, attach hardware to furniture, and secure hinges and other types of fasteners. In addition to woodworking, screws are also used in many other applications, including construction, automotive, and manufacturing. For example, machine screws are commonly used in electronics to hold components in place, while screws are used in construction to fasten drywall and other building materials.

Overall, screws are an important type of simple machine that can be found in many different applications. They allow for a lot of force to be applied to a small area, which makes them useful for holding objects together, lifting objects vertically, and applying pressure to move objects. By understanding how screws work, we can better appreciate the ingenuity of human invention and the ways in which we have been able to harness the power of physics (Hunt, 2003).

7.8. APPLICATIONS OF SIMPLE MACHINES

7.8.1. Construction

Simple machines such as pulleys, levers, and inclined planes are commonly used in construction to lift heavy objects, move materials, and provide mechanical advantage. Cranes are a great example of a machine that uses multiple simple machines. They use pulleys and ropes to lift heavy objects, and wheels and axles to move the crane around the construction site. Bulldozers also use simple machines like levers and hydraulic systems to move and lift heavy materials (Bryan, 1982).

7.8.2. Manufacturing

Simple machines such as screws, gears, and wedges are used in manufacturing to assemble products and machinery. For example, screws are used to hold parts together and can be found in everything from furniture to electronics. Gears are used to transfer power between rotating shafts and are found in machines such as engines and conveyor systems. Wedges are used to cut and shape materials and are commonly found in metalworking and woodworking industries (Wuest et al., 2016).

7.8.3. Transportation

Simple machines such as wheels and axles are used in transportation to reduce friction and increase speed. Vehicles such as cars, bicycles, and trains all utilize wheels and axles to move efficiently and smoothly. In addition to wheels and axles, vehicles also use other simple machines such as levers (for brakes and accelerators) and pulleys (in engines and steering systems) (Bock & Hoberg, 2007).

7.8.4. Agriculture

Simple machines such as pulleys and levers are used in agriculture to help with tasks like lifting and moving heavy loads. For example, pulleys are used in grain elevators to lift and transport grain from one location to another. Levers are used in plows to help turn over soil and in harvesting equipment to cut and gather crops (Jagtap et al., 2022).

7.8.5. Household

Simple machines can also be found in everyday household items. For example, scissors and can openers use wedges to cut through materials. Bottle openers use levers to remove bottle caps. And even something as simple as a doorknob uses a wheel and axle to turn the latch (Venkatesh, 2008).

ACTIVITY 1.1.

Sarah is trying to lift a heavy box onto a shelf in her garage. She knows that she can't lift the box directly, so she decides to use a simple machine to make the task easier.

She looks around and sees a ramp that she could use. If the box weighs 100 pounds and the ramp is 10 feet long and inclined at an angle of 30 degrees, how much force will Sarah need to apply to push the box up the ramp? Assume that the ramp is frictionless, and that Sarah is pushing the box at a constant speed.

7.9. SUMMARY

The chapter on simple machines provides an overview of the different types of simple machines and their applications. It explains how each type of simple machine works, such as how pulleys can be used to lift heavy loads or how inclined planes can be used to reduce the force required to move an object. It also discusses the mechanical advantage of simple machines and how it is calculated. The chapter highlights the importance of simple machines in modern technology, such as in the design of vehicles and machines, and explains the role of simple machines in making work easier. It discusses the advantages and disadvantages of using simple machines, and how efficiency can be improved by reducing friction and other forms of energy loss.

REVIEW QUESTIONS

1. What are the six types of simple machines?
2. How does the mechanical advantage of a simple machine relate to the force required to do work?
3. Give an example of how a pulley can be used to lift a heavy object.
4. How can friction reduce the efficiency of a simple machine?
5. How are simple machines used in modern technology?
6. What is the difference between a wedge and an inclined plane?
7. How can the mechanical advantage of a lever be calculated?
8. What is the difference between a single fixed pulley and a block and tackle pulley system?

MULTIPLE CHOICE QUESTIONS

1. Which of the following is an example of a simple machine?

a. Computer
b. Bicycle
c. Screwdriver
d. Television

2. What is the mechanical advantage of an inclined plane with an angle of 30°?

a. 1
b. 2
c. 0.5
d. 0.6

3. A wheelbarrow is an example of which type of simple machine?

a. Pulley
b. Inclined Plane

c. Lever
d. Wheel and Axle

4. **Which of the following is NOT a type of simple machine?**
a. Screw
b. Wedge
c. Wheel and Axle
d. Gear

5. **What is the formula for calculating work done by a simple machine?**
a. Work = force × distance
b. Work = force / distance
c. Work = distance / force
d. Work = distance × force

Answers to Multiple Choice Questions

1. (c); 2. (c); 3. (c); 4. (d); 5. (d)

REFERENCES

1. Arduini, A., Bussolati, R., Credi, A., Pochini, A., Secchi, A., Silvi, S., & Venturi, M., (2008). Rotaxanes with a calix [6] arene wheel and axles of different length. Synthesis, characterization, and photophysical and electrochemical properties. Tetrahedron, 64(36), 8279–8286.
2. Benjamin, T. B., (1957). Wave formation in laminar flow down an inclined plane. Journal of Fluid Mechanics, 2(6), 554–573.
3. Bitterly, J. G., (1998). Flywheel technology: Past, present, and 21st century projections. IEEE Aerospace and Electronic Systems Magazine, 13(8), 13–16.
4. Bock, S., & Hoberg, K., (2007). Detailed layout planning for irregularly-shaped machines with transportation path design. European Journal of Operational Research, 177(2), 693–718.
5. Brühwiler, E., & Wittmann, F. H., (1990). The wedge splitting test, a new method of performing stable fracture mechanics tests. Engineering Fracture Mechanics, 35(1–3), 117–125.
6. Bryan, J. B., (1982). A simple method for testing measuring machines and machine tools. Part 2: Construction details. Precision Engineering, 4(3), 125–138.
7. Chickering, A. W., & Ehrmann, S. C., (1996). Implementing the seven principles: Technology as lever. AAHE Bulletin, 49, 3–6.

8. Cognard, J., (1986). The mechanics of the wedge test. The Journal of Adhesion, 20(1), 1–13.
9. Cumberbatch, E., (1960). The impact of a water wedge on a wall. Journal of Fluid Mechanics, 7(3), 353–374.
10. Demer, J. L., (2002). The orbital pulley system: A revolution in concepts of orbital anatomy. Annals of the New York Academy of Sciences, 956(1), 17–32.
11. DeRosier, D. J., (1998). The turn of the screw: The bacterial flagellar motor. Cell, 93(1), 17–20.
12. Geeves, M. A., (2002). Stretching the lever-arm theory. Nature, 415(6868), 129–131.
13. Hunt, K. H., (2003). Don't cross-thread the screw! Journal of Robotic Systems, 20(7), 317–339.
14. Insall, R. H., & Machesky, L. M., (2009). Actin dynamics at the leading edge: From simple machinery to complex networks. Developmental Cell, 17(3), 310–322.
15. Jagtap, S. T., Phasinam, K., Kassanuk, T., Jha, S. S., Ghosh, T., & Thakar, C. M., (2022). Towards application of various machine learning techniques in agriculture. Materials Today: Proceedings, 51, 793–797.
16. Kong, L., & Parker, R. G., (2005). Steady mechanics of belt-pulley systems. J. Appl. Mech., 72(1), 25–34.
17. Lister, J. R., (1992). Viscous flows down an inclined plane from point and line sources. Journal of Fluid Mechanics, 242, 631–653.
18. Marulcu, I., & Barnett, M., (2016). Impact of an engineering design-based curriculum compared to an inquiry-based curriculum on fifth graders' content learning of simple machines. Research in Science & Technological Education, 34(1), 85–104.
19. Nelson, A. H., (1980). Mechanical wheels of fortune, 1100–1547. Journal of the Warburg and Courtauld Institutes, 43(1), 227–233.
20. Norbury, J. W., (2006). Working with simple machines. Physics Education, 41(6), 546.
21. Petrescu, R. V., Aversa, R., Akash, B., Bucinell, R., Corchado, J., Apicella, A., & Petrescu, F. I., (2017). Gears-part I. American Journal of Engineering and Applied Sciences, 10(2), 457–472.
22. Pumir, A., Manneville, P., & Pomeau, Y., (1983). On solitary waves running down an inclined plane. Journal of Fluid Mechanics, 135, 27–50.
23. Purcell, T. J., Morris, C., Spudich, J. A., & Sweeney, H. L., (2002). Role of the lever arm in the processive stepping of myosin V. Proceedings of the National Academy of Sciences, 99(22), 14159–14164.
24. Rorres, C., (2000). The turn of the screw: Optimal design of an Archimedes screw. Journal of Hydraulic Engineering, 126(1), 72–80.
25. Sadek, K. S. H., & Daadbin, A., (1990). Improved cam profiles for high-speed machinery using polynomial curve fitting. Proceedings of the Institution of Mechanical Engineers, Part E: Journal of Process Mechanical Engineering, 204(2), 127–132.

26. Sellars, W., (1981). I: The lever of Archimedes. The Monist, 64(1), 3–36.
27. Singh, A., Kumar, A., Chaudhary, R., & Singh, R. C., (2014). Study of 4-wheel steering systems to reduce turning radius and increase stability. In: International Conference of Advance Research and Innovation (ICARI-2014) (Vol. 1, pp. 96–102).
28. Soldatov, D. V., (2006). Soft organic and metal-organic frameworks with porous architecture: From wheel-and-axle to ladder-and-platform design of host molecules. Journal of Chemical Crystallography, 36, 747–768.
29. Thordarson, D. B., Kumar, P. J., Hedman, T. P., & Ebramzadeh, E., (1997). Effect of partial versus complete plantar fasciotomy on the windlass mechanism. Foot & Ankle International, 18(1), 16–20.
30. Veeger, H. E. J., Van, D. W. L. H. V., & Rozendal, R. H., (1991). Load on the upper extremity in manual wheelchair propulsion. Journal of Electromyography and Kinesiology, 1(4), 270–280.
31. Venkatesh, A., (2008). Digital home technologies and transformation of households. Information Systems Frontiers, 10, 391–395.
32. Vowles, H. P., (1932). Early evolution of power engineering. ISIS, 17(2), 412–420.
33. Wartman, J., Bray, J. D., & Seed, R. B., (2003). Inclined plane studies of the Newmark sliding block procedure. Journal of Geotechnical and Geoenvironmental Engineering, 129(8), 673–684.
34. Wuest, T., Weimer, D., Irgens, C., & Thoben, K. D., (2016). Machine learning in manufacturing: Advantages, challenges, and applications. Production & Manufacturing Research, 4(1), 23–45.
35. Yih, C. S., (1963). Stability of liquid flow down an inclined plane. The Physics of Fluids, 6(3), 321–334.
36. Zhu, C., Liu, H., Tian, J., Xiao, Q., & Du, X., (2010). Experimental investigation on the efficiency of the pulley-drive CVT. International Journal of Automotive Technology, 11, 257–261.

CHAPTER 8

APPLICATIONS OF ENGINEERING MECHANICS

UNIT INTRODUCTION

The science of how bodies respond to pressures is the focus of the engineering discipline of engineering mechanics. It is a crucial discipline that serves as the foundation for various engineering specialties and has a broad application spectrum. Engineering mechanics is used in the following situations (Muvdi & McNabb, 2012):

The study of how structures respond to loads is known as structural analysis. Engineering mechanics is essential in this subject as it measures systems' durability, rigidity, and integrity, including bridges, dams, and buildings. The analysis and design of machines rely greatly on engineering mechanics. It assesses how strong and long-lasting machine parts like bearings, gears, and shafts are (Haber, 1990).

Robotics is the science of robots, and engineering mechanics is vital in this subject. The goal of developing and operating robots helps to determine the kinematics and dynamics of the machines. Aerospace engineering is concerned with creating, assembling, and maintaining airplanes and spacecraft. A key technique in this area is engineering mechanics, which determines the aerodynamic characteristics and structural soundness of airplanes and a space probe (Han et al., 2001).

The science of developing and creating vehicles, such as motorcycles, trucks, and cars, is known as automotive engineering. Engineering mechanics is important in this industry since it helps assess the vehicles' efficiency and security. The science of the metabolism of organisms is known as biomechanics. Engineering mechanics is crucial in this subject since it contributes to comprehending the mechanisms of other living species, bones, joints, and mechanics of human movement (Saha et al., 2018).

Engineering mechanics is a fundamental key in the study of materials, which is done through the discipline of materials science. It serves in the design and evaluation of advanced materials and the comprehension of the response of materials under various loads.

In short, engineering mechanics is a basic area that is essential to many engineering professions. It serves various purposes and provides the basis for numerous engineering specialties (Pytel & Kiusalaas, 2016).

Learning Objectives

At the end of this chapter, reader will be able to understand

1. The influence of loads on different types of structures.
2. The analytical methods used in structural analysis and their limitations.
3. The effectiveness of materials and methods used in structural analysis.
4. The application of principles of machine design in designing and analyzing mechanical components.
5. The concept of robotics and its various applications.
6. The principles of aerospace engineering and its applications.
7. The principles of automotive engineering and their applications.
8. The principles of biomechanics and their applications.
9. The role of materials engineering in modern engineering applications.

Key Terms

1. Structural Analysis
2. Analytical Methods
3. Machine Design
4. Robotics
5. Aerospace Engineering
6. Automotive Engineering
7. Biomechanics
8. Materials Engineering

8.1. STRUCTURAL ANALYSIS

An area of solid mechanics called structural analysis helps engineers make decisions by using simpler models of solids like beams, shells, and bars. Its primary goal is to ascertain how loads affect the physical structures and the parts they compose. Unlike the theory of elasticity, differential equations in one spatial variable are frequently used in structural analysis models. Any materials that would endure loads, like ships, aircraft, buildings, and bridges, are subjected to this kind of examination. Structural analysis computes a structure's deformations, stress distribution, loads, support responses, accelerations, stability, and velocity, using concepts from applied mathematics, applied mechanics, and materials science. The analysis confirms a structure's suitability for usage, frequently displacing physical examinations. Thus, structural analysis is a crucial aspect of the engineering design of structures (Anderson et al., 2004) (Figure 8.1).

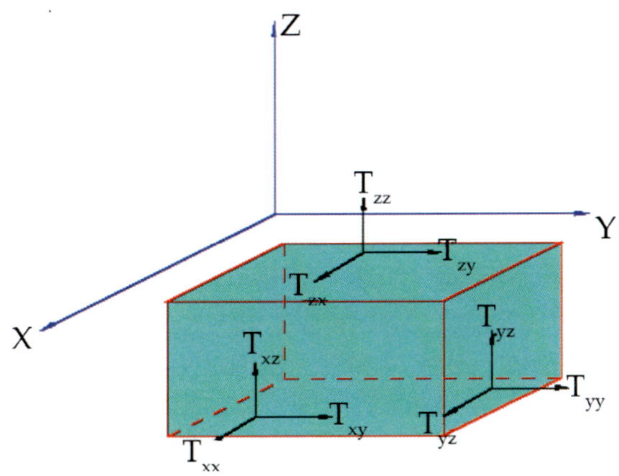

Figure 8.1. Illustration of theory of elasticity.

Source: S. Dhar, Creative Commons License.

8.1.1. Influence of Loads on Structures

A structure is a body or a network of interconnected pieces supporting a structural analysis load. Bridges, towers, and buildings are significant examples of civil engineering; mechanical systems, tanks, pressure vessels, ship and aircraft frames, and electrical structures are significant examples in other engineering areas. An engineer must consider a structure's workability, esthetic appeal, and safety when designing it, as well as its cost and environmental impact. Other engineering specialties deal with a wide range of non-building structures (Ganapathi et al., 2006).

8.1.2. Types of Structures

A structural system comprises structural components and the materials they are made of. A structural engineer must be competent in identifying a structure's various components and categorizing it according to its design or use. In addition to connecting rods, trusses, beams, and columns, structural elements that direct systemic forces through materials include cables, arches, cavities, channels, and even angles, surface structures, and frames (Astroza et al., 2015).

8.1.3. Loads on Different Structures

It is vital to identify the loads the structure should support after establishing dimensional requirements for the structure. Hence, describing the loads acting on the structure is the first step in structural design. Building standards frequently specify a structure's design loading. General construction regulations and design standards are two different categories, and engineers must adhere to all of the code's standards for the structure to remain effective (Lee & Fenves, 1998).

Remember

Consider the different types of loads that can act on structures such as tension, compression, bending, and torsion. It is important to understand how these loads affect the structure and design appropriate support and reinforcement to ensure its stability and safety.

Structure engineering must deal with two different sorts of loads when designing. The initial sort of load is dead loads, which are made up of the masses of the different structural components and any items that are affixed to the building constantly. For instance, the walls, electrical fittings, windows, floor slab, plumbing, and other auxiliary components. Live loads differ in size and position and are the second load category. Live loads come in various forms, including those associated with structures, bridges, collisions, wind, earthquakes, snow, and other natural stresses (Bernal, 2002).

8.1.4. Analytical Methods in Structural Analysis

To conduct an appropriate analysis, a structural engineer must ascertain data such as geometry, design integrity, and materials' qualities. Support responses, pressures, and deflections are often included in this study. Then, this data is contrasted with standards that specify the failure conditions. Innovative structural analysis may look at durability, non-linear, and dynamic response. The analysis can be approached from three angles: the finite element approach, the elasticity theory approach, and the mechanics of the material approach (Dinev, 2012). The first two involve analytical equations that typically use basic linear elastic concepts, resulting in restricted solutions that can frequently be solved manually. The finite element approach is a numerical method for resolving differential equations produced by principles of mechanics like the elasticity theory and strength of materials. The finite-element approach is more appropriate for structures of any complexity or size and highly dependent on the computational capacity.

The concept is built on the same three essential relations, compatibility, constitutive, and equilibrium, regardless of methodology (Basseville et al., 2010) (Figure 8.2).

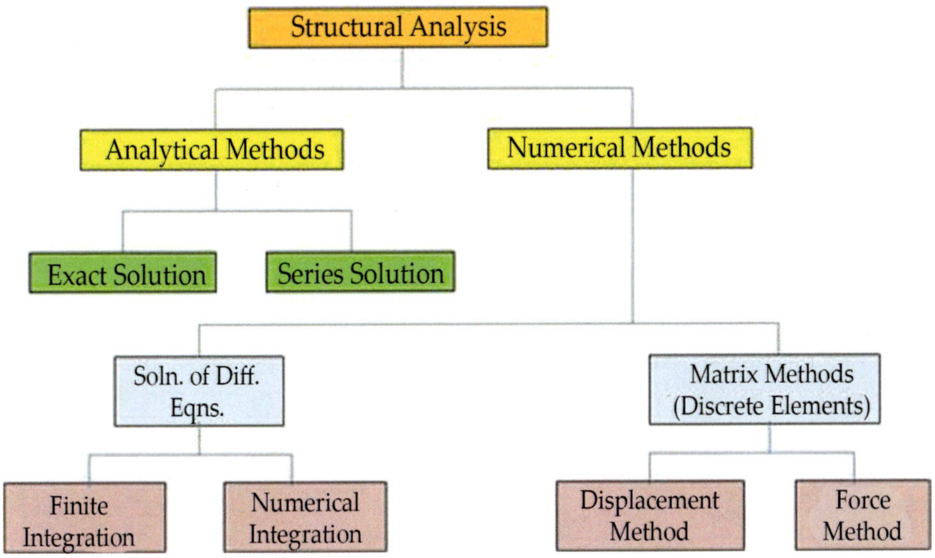

Figure 8.2. Flow diagram of branches of structural analysis methods.

Source: Fem Best Practices, Creative Commons License.

8.1.5. Limitations in SA

Each technique has notable drawbacks. Applying the materials mechanics method is restricted to basic structural components under basic loading circumstances. The approved structural components and loading scenarios adequately address many practical engineering issues. In theory, structural elements with general geometry and general loading conditions can be solved using the theory of elasticity. Moreover, analytical solutions are only applicable to comparatively basic instances. To solve an elasticity problem, one must additionally solve a system of partial differential equations (PDEs), which is a far more difficult mathematical task than solving a problem in the mechanics of materials that only calls for solving an ordinary differential equation (ODE). The finite element approach is maybe the most conservative while also being the most effective. This approach depends on other structural theories (like those listed here) to solve the equations. With the caveat that there is some numerical inaccuracy, this makes it possible to solve these equations, even with extremely complicated geometry and loading strategy. This strategy must be used with a clear grasp of its limits to be effective and accurate (Vořechovský & Novák, 2009).

8.1.6. Effectiveness of Materials Methods

The mechanics of materials method is the simplest of the three approaches we have explored here. It can be used for basic structural elements subject to particular loads, such as colorful beams in pure bending, axially loaded bars, and circular shafts subjected to bending. The principle of superposition can be used to examine a member subjected to combined loads under certain circumstances by superimposing the solutions. For typical constructions like thin-walled pressure vessels, there are solutions for special instances (Erofeev et al., 2016).

TIPS

"Consider all factors when designing a machine, including material selection, operating conditions, safety, and efficiency, to ensure optimal performance and longevity."

This strategy can be combined with statics to analyze systems, leading to the methods of segments and junctions for analyzing trusses, the moment distribution method for analyzing small rigid frames, and the portal frame and cantilevered methods for analyzing big rigid frames. These techniques were created in their current iterations during the second part of the 19th century, except for moment distribution, which started to be used in the 1930s. They are still utilized for tiny structures and major structure prototypes.

The answers have relied on the linear isotropic infinitesimal elasticity and Euler Bernoulli beam theory (Wilson et al., 1988).

8.2. MACHINE DESIGN

Machine Design or Mechanical Design is one of the key fields of Engineering Design. Let's use the example of a car's gearbox to clarify the exact machine design. The gearbox helps the engine to supply motion and energy to the vehicle's wheels. The gearbox comprises several gears that must endure both the power and mobility of the vehicle (Ugural, 2018). The gears must be built to function at the correct speeds and support the necessary loads. The gear of a certain material and precise dimensions that could bear all loads and be built at the least possible cost offering the best efficiency is designed after numerous calculations are performed, taking required speeds and loads into account. Similarly, every part of the car, such as the engine, must be constructed to satisfy all functional criteria as efficiently as feasible. The term Machine Design or Mechanical Design refers to the entire design process (Childs, 2013) (Figure 8.3).

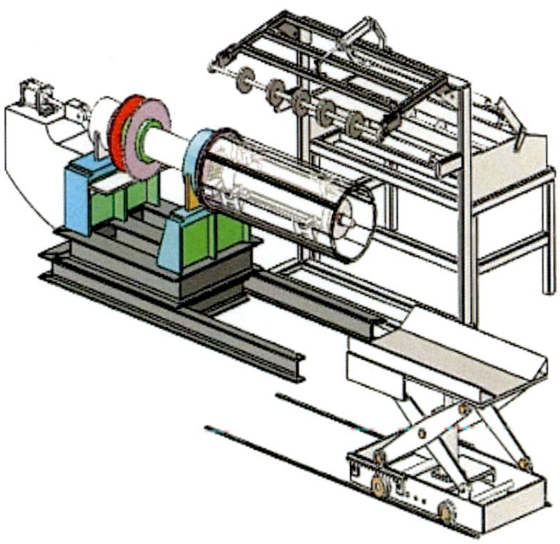

Figure 8.3. Image of machine design.

Source: Rabik, Creative Commons License.

The process by which resources are transformed into mechanical work, or the processes to find the valuable result from the machines in the intended shape following the needs of humans, is known as mechanical design or machine design. A new machine may be developed wholly due to machine design, or a current machine may be upgraded or improved. For example, a new gearbox can be designed if the current gearbox is excessively heavy or is unable to support the actual loads. Yet, if the same gearbox can lift larger loads, it can be improved by making significant design adjustments (Cruse, 1997).

8.3. CONCEPT OF ROBOTICS

Robotics is the interdisciplinary branch of technology that deals with the design, construction, operation and application of robots. It aims to develop machines that can replicate human actions. Mechanical engineering is a significant branch in engineering, having to its credit the design and uses of some wonders such as aeroplanes, ships, automobiles, and automation. Robotics in mechanical engineering encompasses the composition and structure of robots. It aims to design intelligent machines that can help and assist humans in their day-to-day lives and keep them safe when used in dangerous environments (Davidson et al., 2004).

Robots are useful in high-risk situations, such as bomb detection and deactivation, production, and radioactive material inspection procedures. Robots can be created in any shape or size. Yet, bio-inspired models can mimic human behavior and are created to seem like human situations. They can mimic human speech, thought, lifting, walking, and other activities. While certain types of robots work autonomously, others may need user involvement (Berenguel et al., 2016).

An extensive range of applications exists for robotics engineering. Robots are utilized to manufacture consumer and industrial goods and for production, assembling, packaging, transportation in space and on Earth, warfare, surgery, and research. Robots may learn to perform activities more precisely, consistently, and affordably than people if integrated with IA (Ben-Ari et al., 2018) (Figure 8.4).

Figure 8.4. Image of robot.

Source: The Big Red Group, Creative Commons License.

8.4. AEROSPACE ENGINEERING

An engineering area that concerns the design, production, evaluation, and execution of vehicles that function in the environment in space or on Earth is known as aerospace engineering, also called astronautical engineering or aeronautical engineering. In 1958, the first description of aeronautical engineering appeared, treating the

Earth's atmosphere and the space above it as a particular domain for developing flight apparatus. Nowadays, the term "aerospace" is widely used in place of the terms "astronautical engineering" and "aeronautical engineering" (Brodeur et al., 2002).

The development of a flight vehicle requires knowledge of several different engineering expertise. Rarely can one individual address everything; instead, most companies have engineering teams with specialists in electronics, materials science, launch vehicles, aerodynamics, structural design, and stabilization and management systems. Several designs are available that consider the vehicle's needs, the underlying infrastructure, and the project's financial viability (Nunes & Silva, 2016).

Did you Know?

Did you know that the Apollo Lunar Module, which carried astronauts to the moon during the Apollo missions, was designed using principles of aerospace engineering, including aerodynamics, structural analysis, and materials engineering?

8.5. AUTOMOTIVE ENGINEERING

Since the widespread use of cars made to carry people, automobile engineering has grown in relevance and importance. There is a big need for automotive engineers because the auto industry and auto component companies are growing quickly. One of the most difficult and varied careers is automobile engineering, sometimes known as vehicle or automotive engineering (Bewilogua et al., 2009).

This sector focuses on producing, testing, maintaining, developing, and repairing vehicles such as scooters, trucks, cars, motorcycles, and other types of vehicles and the associated sub-engineering systems. To design the required combination for producing and developing automobiles, automobile engineering integrates the features of numerous engineering elements, including electronic, mechanical, computer, electrical, and safety engineering (Modrea et al., 2013).

8.6. BIOMECHANICS

The study of biological systems, specifically their forms and function, is known as biomechanics. This field of research uses techniques from mechanics focused on the impact of forces on the body's movement. At least since the Renaissance, when Italian physician and physicist Giovanni Alfonso Borelli first outlined the fundamentals of muscle and skeletal dynamics, ideas, and research into biomechanics have been pursued. In the 20th century, biomechanics research gained increasing recognition (Chao et al., 1994).

Modern biomechanics is an interdisciplinary field that integrates information from the medical and biological sciences with experience in engineering sciences. Cell biomechanics, Cardiovascular biomechanics, human movement biomechanics, occupational biomechanics, and sport biomechanics are just a few of the many subspecialties in biomechanics. Sports biomechanics is one area where it works to help athletes function better and avoid injuries. Biomechanical analysis is employed in occupational biomechanics to comprehend and improve the mechanical relationship of workers with their surroundings (Knudson & Knudson, 2007).

Daily lives are impacted by many of the inventions that emerged out of biomechanics research. For instance, the emergence of labor biomechanics prioritizes boosting worker productivity without compromising workplace safety. It led to the creation of modern furniture, equipment, and other workplace fixtures that lessen the physical strain on employees' bodies. Another advancement is clinical biomechanics, which interprets and analyzes typical and unusual human physiology and anatomy using mechanical facts, techniques, and mathematics (Robertson et al., 2013).

The construction of prosthetic limbs for amputee veterans received much attention during World Wars (I & II), which greatly advanced biomechanics and rehabilitation therapy. That research aimed to make orthopedic implants, like those used to transplant the knee or hip, more mechanically effective. Biomechanics study also contributed to advancements in walking aids for cerebral palsy patients and people with lower-leg amputations. With the advent of a novel prosthetic foot category that stores and releases mechanical energy while walking, amputees' metabolic costs may be reduced, and people with amputations can now participate in sports. People with impairments could increase their mobility because of the ergonomic design of environmental features like stairs and the physiologically sound design of adaptive equipment like wheelchairs (Humphrey & Delange, 2004).

Biomechanics has a broad spectrum of uses. The design of surgically implanted prostheses, like heart transplants and small blood vessels, the technology of body tissue, like heart valves and intervertebral discs, and prevention of injury associated with automobile accidents, such as low-speed accidents like small-scale soft-tissue injuries and high-speed accidents like fatal injuries, are a few case studies of how biomechanical assessment is used (Özkaya et al., 2012).

8.7. MATERIALS ENGINEERING

Numerical methods are used to analyze ductile solids in materials mechanics, a subfield of mechanical engineering. Resistance is the propensity of an element to withstand applied efforts and forces without breaking, enduring deformation, or acquiring degradation (Hubbell et al., 2009).

The material's durability is frequently inadequate for mechanical design components that have intricate, necessitating the employment of methods that utilize the concepts of elasticity or the mechanics of malleable solids. Using numerical techniques like finite element analysis can then address these issues with respect to stresses and strains (Ling et al., 2018).

ACTIVITY 8.1.

How does the study of biomechanics help in designing prosthetic limbs for amputees?

8.8. SUMMARY

The chapter on Applications of Engineering Mechanics covers various fields where the principles of mechanics are applied, including Structural Analysis, Machine Design, Robotics, Aerospace Engineering, Automotive Engineering, Biomechanics, and Materials Engineering. Structural Analysis involves analyzing the behavior of structures under various loads and stresses. This includes calculating forces and stresses in different structural components such as beams, trusses, and frames. Machine Design is the process of designing machines that can perform various functions, such as manufacturing, transportation, and energy conversion. This involves applying the principles of mechanics to design and analyze the components and mechanisms of machines. Robotics is an interdisciplinary field that combines principles from mechanics, electronics, and computer science to design and control robots. Robotics applications include manufacturing, space exploration, and medical devices. Aerospace Engineering involves designing and developing aircraft, spacecraft, and related components. This field uses mechanics principles to analyze the behavior of aircraft and spacecraft under various conditions, including aerodynamics and structural loads. Automotive Engineering applies mechanics principles to the design and development of automobiles and other vehicles. This includes analyzing the performance and safety of vehicles under different conditions, such as speed and terrain. Biomechanics is the study of the mechanics of biological systems, including humans and animals. This field uses mechanics principles to analyze the behavior of biological systems and to design medical devices and prosthetics. Materials Engineering involves designing and developing materials with specific mechanical properties for various applications. This field applies mechanics principles to understand the behavior of materials under different conditions and to optimize their properties.

REVIEW QUESTIONS

1. What is the importance of structural analysis in engineering?
2. What are the principles of machine design?
3. What are the different applications of robotics in the engineering field?
4. What is the importance of aerospace engineering in the field of engineering?
5. What are the principles of biomechanics and how are they applied in the engineering field?

MULTIPLE CHOICE QUESTIONS

1. What is the concept of structural analysis?

a. The study of materials
b. The study of structures
c. The study of machines
d. The study of robotics

2. What is machine design?

a. The design of structures
b. The design of machines
c. The design of robots
d. The design of aircraft

3. What is the concept of robotics?

a. The study of materials
b. The study of structures
c. The study of machines
d. The study of robots

4. What is aerospace engineering?

a. The study of aircraft and spacecraft
b. The study of machines
c. The study of materials
d. The study of robots

5. What are biomechanics?

a. The study of machines
b. The study of robots
c. The study of materials
d. The study of the mechanics of living organisms

Answers to Multiple Choice Questions

1. (b) 2. (b) 3. (d) 4. (a) 5. (d)

REFERENCES

1. Anderson, K., El-Sheikh, A., & Newson, T., (2004). Application of structural analysis to the mechanical behavior of the cornea. Journal of the Royal Society Interface, 1(1), 3–15.
2. Astroza, R., Ebrahimian, H., & Conte, J. P., (2015). Material parameter identification in distributed plasticity FE models of frame-type structures using nonlinear stochastic filtering. Journal of Engineering Mechanics, 141(5), 04014149.
3. Basseville, M., Bourquin, F., Mevel, L., Nasser, H., & Treyssède, F., (2010). Handling the temperature effect in vibration monitoring: Two subspace-based analytical approaches. Journal of Engineering Mechanics, 136(3), 367–378.
4. Ben-Ari, M., Mondada, F., Ben-Ari, M., & Mondada, F., (2018). Robots and their applications. Elements of Robotics, 1–20.
5. Berenguel, M., Rodríguez, F., Moreno, J. C., Guzmán, J. L., & González, R., (2016). Tools and methodologies for teaching robotics in computer science & engineering studies. Computer Applications in Engineering Education, 24(2), 202–214.
6. Bernal, D., (2002). Load vectors for damage localization. Journal of Engineering Mechanics, 128(1), 7–14.
7. Bewilogua, K., Bräuer, G., Dietz, A., Gäbler, J., Goch, G., Karpuschewski, B., & Szyszka, B., (2009). Surface technology for automotive engineering. CIRP Annals, 58(2), 608–627.
8. Brodeur, D. R., Young, P. W., & Blair, K. B., (2002). Problem-based learning in aerospace engineering education. In: Proceedings of the 2002 American Society for Engineering Education Annual Conference and Exposition, Montreal, Canada (pp. 16–19).
9. Chao, E. Y., Neluheni, E. V., Hsu, R. W., & Paley, D., (1994). Biomechanics of malalignment. The Orthopedic Clinics of North America, 25(3), 379–386.
10. Childs, P. R., (2013). Mechanical Design Engineering Handbook (Vol. 1, pp. 1–10). Butterworth-Heinemann.
11. Cruse, T. A., (1997). Reliability-Based Mechanical Design (Vol. 1, 108, pp. 1–12). CRC press.
12. Davidson, J. K., Hunt, K. H., & Pennock, G. R., (2004). Robots and screw theory: Applications of kinematics and statics to robotics. J. Mech. Des., 126(4), 763, 764.
13. Dinev, D., (2012). Analytical solution of beam on elastic foundation by singularity functions. Engineering Mechanics, 19(6), 381–392.

14. Erofeev, V., Dergunova, A., Piksaikina, A., Bogatov, A., Kablov, E., Startsev, O., & Matvievskiy, A., (2016). The effectiveness of materials different with regard to increasing the durability. In: MATEC Web of Conferences (Vol. 73, p. 04021). EDP Sciences.
15. Ganapathi, M., Prakash, T., & Sundararajan, N., (2006). Influence of functionally graded material on buckling of skew plates under mechanical loads. Journal of Engineering Mechanics, 132(8), 902–905.
16. Haber, R. B., (1990). Visualization techniques for engineering mechanics. Computing Systems in Engineering, 1(1), 37–50.
17. Han, B., Post, D., & Ifju, P., (2001). Moiré interferometry for engineering mechanics: Current practices and future developments. The Journal of Strain Analysis for Engineering Design, 36(1), 101–117.
18. Hubbell, J. A., Thomas, S. N., & Swartz, M. A., (2009). Materials engineering for immunomodulation. Nature, 462(7272), 449–460.
19. Humphrey, J. D., & Delange, S. L., (2004). An introduction to biomechanics. Solids and Fluids, Analysis and Design (Vol. 1, pp. 1–15). Springer, Heidelberg.
20. Knudson, D. V., & Knudson, D., (2007). Fundamentals of Biomechanics (Vol. 183). New York: Springer.
21. Lee, J., & Fenves, G. L., (1998). Plastic-damage model for cyclic loading of concrete structures. Journal of Engineering Mechanics, 124(8), 892–900.
22. Ling, S., Kaplan, D. L., & Buehler, M. J., (2018). Nanofibrils in nature and materials engineering. Nature Reviews Materials, 3(4), 1–15.
23. Modrea, A., Vlase, S., Teodorescu-Draghicescu, H., Mihalcica, M., Calin, M. R., & Astalos, C., (2013). Properties of advanced new materials used in automotive engineering. Optoelectronics and Advanced Materials-Rapid Communications, 7, 452–455.
24. Muvdi, B. B., & McNabb, J. W., (2012). Engineering Mechanics of Materials (Vol. 1, pp. 1–22). Springer Science & Business Media.
25. Nunes, J. P., & Silva, J. F., (2016). Sandwiched composites in aerospace engineering. In: Advanced Composite Materials for Aerospace Engineering (pp. 129–174). Woodhead Publishing.
26. Özkaya, N., Nordin, M., Goldsheyder, D., & Leger, D., (2012). Fundamentals of Biomechanics (Vol. 86). New York: Springer.
27. Pytel, A., & Kiusalaas, J., (2016). Engineering Mechanics: Dynamics (Vol. 1, pp. 1–5). Cengage Learning.
28. Robertson, D. G. E., Caldwell, G. E., Hamill, J., Kamen, G., & Whittlesey, S., (2013). Research Methods in Biomechanics (Vol. 1, pp. 1–20). Human kinetics.
29. Saha, S., Mishra, M. K., Reddy, C. M., & Desiraju, G. R., (2018). From molecules to interactions to crystal engineering: Mechanical properties of organic solids. Accounts of Chemical Research, 51(11), 2957–2967.

30. Ugural, A. C., (2018). Mechanical Design of Machine Components: SI Version (Vol. 1, pp. 1–11). Taylor & Francis.
31. Vořechovský, M., & Novák, D., (2009). Correlation control in small-sample Monte Carlo type simulations I: A simulated annealing approach. Probabilistic Engineering Mechanics, 24(3), 452–462.
32. Wilson, P. F., Ma, M. T., & Adams, J. W., (1988). Techniques for measuring the electromagnetic shielding effectiveness of materials. I. Far-field source simulation. IEEE Transactions on Electromagnetic Compatibility, 30(3), 239–250.

INDEX

E

F

T

U

V

W

Y

Kolibri 360 GmbH
Ferdinand-Jühlke-Straße 7
99095 Erfurt, Deutschland
produktsicherheit@kolibri360.de